Essentials of Automatic Control with MATLAB® in 20 Lessons

This book is designed to explain the fundamental principles of automatic control through 20 lessons, each incorporating worked examples and MATLAB®-based exercises to help readers effectively understand and apply the proposed methods. It offers a concise and accessible resource for learning automatic control, blending simplicity with clarity. Drawing on decades of teaching experience, the authors aim to provide an approachable introduction to the core concepts of the discipline, encouraging further exploration through independent study.

Primarily intended for undergraduate engineering students, the book is also suitable for a wider audience. As automatic control is a foundational subject across numerous academic programs, this resource equips readers with essential analytical tools and introduces key problems, fostering deeper insights into the subject. The integration of MATLAB® within the 20-lesson framework enhances learning by supporting both theoretical understanding and practical problem-solving.

This book champions the value of traditional learning approaches while promoting critical thinking and problem-solving skills, essential competencies in today's AI-driven world.

The primary focus is on classical topics related to Single-Input Single-Output (SISO) linear continuous-time systems. Additionally, the book provides introductory material on topics such as Multiple-Input Multiple-Output (MIMO) and discrete-time systems, serving as a bridge to more advanced studies. By establishing a strong foundation in these areas, the book prepares readers to tackle complex challenges in modern automatic control and excel in future academic and professional endeavors.

Luigi Fortuna earned a degree in electrical engineering (cum laude) from the University of Catania, Italy, in 1977. He is a Full Professor of System Theory at the University of Catania. From 2005 to 2012, he was the Dean of the Engineering Faculty. He has been a visiting researcher at the Joint European Torus in Abingdon UK. He currently teaches complex adaptive systems and robust control. He has published 14 scientific books and 12 industrial patents. Dr. Fortuna has been a consultant of STMicroelectronics and other companies. He is the Editor in Chief of the Springer-Briefs Series on Nonlinear Circuits. His scientific interests include robust control, Tokamak machine control, complex engineering, nonlinear circuits, chaos, cellular neural networks, robotics, and smart devices for control. Additionally, he is an IEEE Fellow.

Mattia Frasca graduated in Electronics Engineering in 2000 and earned his PhD in Electronics and Automation Engineering in 2003, from the University of Catania, Italy. Currently, he is a research associate at the University of Catania, where he also teaches process control. His scientific interests include nonlinear systems and chaos, complex networks, and bio-inspired robotics. He is Associate Editor of the *International Journal of Bifurcations and Chaos*, and Editor of *Chaos, Solitons and Fractals.* He is the general chair for the next edition of the European Conference on Circuits Theory and Design to be held in Catania. Additionally, he is co-author of one research monograph with Springer, three with World Scientific, and one book on Optimal and robust control with CRC Press. Dr. Frasca has published more than 250 papers on refereed international journals and international conference proceedings and is co-author of two international patents. He is an IEEE Senior and a member of the Board of the Italian Society for Chaos and Complexity (SICC).

Arturo Buscarino graduated in Computer Science Engineering in 2004 and earned his PhD in Electronics and Automation Engineering in 2008, from the University of Catania, Italy. Currently, he is a Technical Assistant at the University of Catania and teaches Modeling and Optimization at the Laura Magistrale in Management Engineering. He collaborates with the EUROFusion Consortium, ENEA Frascati, and JET Culham, UK. Dr. Buscarino has been a visiting researcher at the University of Wisconsin-Madison, US. His scientific interests include nonlinear circuits and systems, chaos and synchronization, complex networks, control systems, Cellular Nonlinear Networks, and plasma engineering. He is Associate Editor of *Cogent Engineering.* Dr. Buscarino has published one research monography on nonlinear circuits, and more than 60 papers on refereed international journals and international conference proceedings.

Essentials of Automatic Control with MATLAB® in 20 Lessons

Luigi Fortuna, Mattia Frasca and
Arturo Buscarino

CRC Press
Taylor & Francis Group
Boca Raton London New York

CRC Press is an imprint of the
Taylor & Francis Group, an **Informa** business

First edition published 2026
by CRC Press
2385 NW Executive Center Drive, Suite 320, Boca Raton FL 33431

and by CRC Press
4 Park Square, Milton Park, Abingdon, Oxon, OX14 4RN

CRC Press is an imprint of Taylor & Francis Group, LLC

ISBN: 978-1-032-78303-1(hbk)
ISBN: 978-1-032-78308-6 (pbk)
ISBN: 978-1-003-48728-9 (ebk)

DOI: 10.1201/9781003487289

Typeset in CMR10 font
by KnowledgeWorks Global Ltd.

Dedicated to our wives

Contents

Preface

Today, there is a significant number of books on automatic control available for undergraduate and graduate students, as well as researchers in the field. Despite the abundance of these resources, our aim is to offer a book that takes a different approach, guiding readers through the core topics of the discipline in a concise format, enriched with numerous worked examples.

This book is, in fact, conceived to explain the fundamental principles of automatic control through 20 lessons. The aim is to provide a concise, essential, and self-contained resource to help students grasp the principles of the discipline. To follow the content of the book, only basic knowledge of calculus, algebra, and physics is required.

The book is primarily intended for bachelor students in engineering, but it is not restricted to them. As automatic control is a fundamental subject in many academic programs, the book is designed to equip students with both the essential analytical tools and key problems, whose solutions foster a deeper understanding of the discipline. The readers will benefit from the use of MATLAB®, which is integrated into the 20-lesson course to support learning and problem-solving throughout the book.

The motivation behind this book lies in the need for a resource in automatic control that combines both a concise style and an accessible, straightforward approach. Drawing on decades of experience, the authors aim to provide a clear and simple introduction to the fundamental principles of the discipline, encouraging readers to further develop their understanding through additional studies.

The structure of the book allows readers to gain a solid understanding of the key concepts in a relatively short amount of time, motivating further exploration of the discipline. In the era of AI, a book like this remains valuable because it preserves the historical context and foundational principles of automatic control. While technology evolves rapidly, understanding classical methods provides a strong foundation for future advancements. This book encourages the new generation to appreciate traditional learning approaches, fostering critical thinking and problem-solving skills that are essential in an AI-driven world.

The book primarily focuses on classical subjects related to SISO (Single-Input Single-Output) linear continuous-time systems. However, it also introduces additional topics that can serve as a gateway to understanding MIMO (Multiple-Input Multiple-Output) and discrete-time systems. Instead, nonlinear systems are not covered in this textbook. The book lays down fundamental principles that will enable readers to explore advanced topics in automatic control, such as optimal control and robust control. By establishing a solid foundation, the book prepares readers to tackle more complex concepts in the field, ensuring they are well-equipped for future studies and applications in modern automatic control. The book is organized to include many fundamental concepts of systems theory, which serve as prerequisites for understanding the main subjects of automatic control.

The authors designed this book to serve as a foundational resource for a course that can be offered before other key subjects such as electrical circuits, electronics, or signal processing. This strategic placement positions the book as a core reference for essential principles in engineering studies, broadening its appeal to a wide range of readers.

The book is structured into 20 lessons, each featuring a variety of worked examples and MATLAB®-based exercises designed to help the reader grasp the proposed methods effectively. The initial lessons introduce the concept of automatic control, focusing on modeling systems through state-space representations and transfer functions. Following this foundation, the course examines system behavior analysis, covering the time response to canonical signals in both first-order and second-order systems, as well as stability analysis and frequency response. Subsequently, the characteristics of closed-loop systems are explored. Performance indicators are introduced, along with criteria for closed-loop stability. The remaining lessons are dedicated to illustrating techniques for controller synthesis, highlighting three main approaches: designing controllers based on frequency response, tuning PID controllers, and addressing the eigenvalue assignment problem in state-space representation.

For MATLAB® and Simulink® product information, please contact:

The MathWorks, Inc.
3 Apple Hill Drive
Natick, MA 01760-2098 USA
Tel: 508-647-7000
Fax: 508-647-7001
Email: info@mathworks.com
Web: www.mathworks.com

1

Lecture 1—Automatic control

In this first lecture we introduce the notion of system, the main idea underlying automatic control and the concept of feedback. We then discuss a few examples and some historical notes on automatic control.

1.1   The notion of control and system

Controlling a process means to force it to follow a desired operation. The process or the system to control can be a plant, a device, or a natural phenomenon. Such a system is, generally, influenced by several quantities, which are called the inputs, and, in turn, it influences the environment through other variables, which are called the outputs, or variables of interest. Schematically this is represented in Fig. 1.1, where the system is shown as a box, the inputs as arrows pointing at it, and the outputs as arrows starting from it.

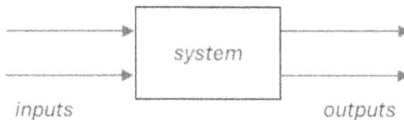

FIGURE 1.1
Block representation of a system.

Here, the direction of the arrow is very important. It represents the direction of the cause-effect relationship. The input can be viewed as the cause of the output, which is the effect. The system is responding to the causes with an effect determined by the specific characteristics of the process itself.

The inputs can be either manipulated variables or disturbances. The first term refers to variables that are used to control the process. The second term refers to variables whose evolution in time is not controlled, but they still influence the system.

Among the outputs of the system, we find the controlled variables, namely those variables whose evolution in time we want to target with the control. Generally, the goal of the control is to ensure that such variables are equal

DOI: 10.1201/9781003487289-1   1 (or at least very close) to the reference signal. In case the reference signal is constant in time, it is usually called set point.

Hence, the control makes it possible to reach some high-level goal, such as safety, or environment protection, or profit, by letting some quantities of interest in the process follow a desired behavior.

1.2   The feedback loop and its constitutive elements

To illustrate how control operates, let us consider two examples from our daily life.

Consider a car being driven by a human on a highway. The goal of the control system is to keep the car within its lane. To achieve this, the driver continuously gathers information about the car's position, primarily using visual input, and adjusts the steering wheel as needed to correct the car's position if it deviates from the desired trajectory. In this example, the system to be controlled is the car, which is influenced by various inputs. These include variables manipulated by the driver, such as the steering angle, as well as external disturbances like irregularities in the road surface, wind, and other factors. The primary output of the system being controlled is the car's position within the lane. However, there are additional outputs, such as the car's velocity, which in this example is not controlled but could be a controlled variable in other scenarios.

The second example is the control of a temperature in a room equipped with a heating element. In this case, the goal of the control is to keep the temperature in a comfortable range. This task is achieved thanks to the presence of a thermostat that senses the temperature of the room and a controller that, based on the measured temperature, switches on and off the heating element (in a first simplified scenario we are here considering that the heating element can only be switched on or off).

Despite the two examples pertaining to different fields, they share many common elements. First of all, in both cases the control aims at a specific goal, which is obtained by steering some variables to desired behaviors. Then, in both cases we find: a sensor that measures the controlled variable (the eyes or the thermostat); an actuator (the steering or the heating element); and a controller (the human brain or the automatic device). Finally, the way the control operates is similar in the two cases: the control aims at reducing the discrepancy between desired and actual value of the controlled variable. In fact, in both cases the control operates in a feedback loop. To illustrate this concept, we refer to Fig. 1.2.

In Fig. 1.2 we find that the output of the system is measured by the sensor. This quantity (the measured output) is then compared with the reference, yielding an error defined as the difference between reference and measured

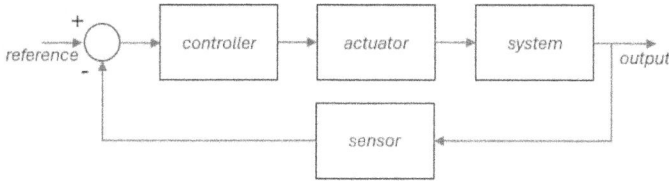

FIGURE 1.2
The feedback loop.

output. This error constitutes the input of the controller. Based on the values of the error, the controller establishes the proper action to do on the process. This is done through the actuator.

The control operates in a loop, as the operations previously described are continuously done. This loop is called the feedback loop, as the core idea is to bring back the output to build the error signal. More specifically, this is a negative feedback scheme, where the target is to reduce the error.

In summary the key elements of a feedback control loop are: the system, which represents the plant to be controlled; the sensor, which provides measurements of the system's output; the actuator, the device supplying the energy required to control the plant; and the controller, which processes the error signal to determine the corrective action. Designing this corrective action is the primary objective of control design.

1.3   Open-loop control

The negative feedback loop is the most common approach in control systems, but it is not the only one. Positive feedback loops also exist, in which the controller input is the sum of the reference signal and the measured output. Additionally, there are scenarios where output information is not fed back to the controller. This is the case with open-loop control, where the system is controlled without using feedback from the output. In such instances, control is achieved by predetermining the input signals to reach the desired outcome. This approach is applied when measurements are impractical or do not yield significant benefits. For successful open-loop control, an accurate model of the system is essential, and the system must already be stable. Although we have yet to introduce the formal concept of stability, it is sufficient for now to consider it as a condition in which the system's behavior remains bounded. The controller that operates in open-loop sends a control signal to the system without receiving information about the result. This is what occurs for instance when a timer turns a light on and off at predefined intervals, or when an oven heats for a set time and then turns off, regardless of the temperature reached.

1.4   Feedforward control

Similarly to open-loop control, also feedforward control is not based on output information. Feedforward control is, for instance, used to compensate for the effect of a disturbance frequently occurring in the system. This disturbance is measured and then the controller uses this information to design a proper action to perform on the system, anticipating the effect of the disturbance. The feedforward control scheme is illustrated in Fig. 1.3. The feedforward control requires an almost perfect knowledge of how the system operates and is often used in synergy with the feedback loop, as shown in Fig. 1.4.

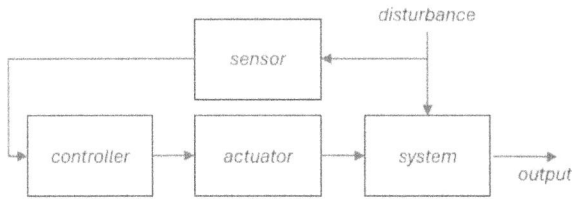

FIGURE 1.3
Scheme of the feedforward control.

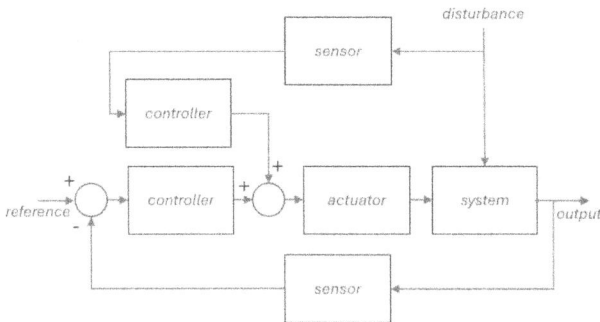

FIGURE 1.4
Scheme of the feedforward control combined with feedback control.

1.5 Manual and automatic control

In the two examples that we have discussed control is operated in a different way. In the car example, calculations on the adjustment of the car trajectory are done by the human brain, whereas in the room temperature example the calculations are done by an electronic device. This is a general feature, and indeed, control can be operated manually or automatically. The two approaches have different benefits and drawbacks. Manual control has the advantage of ensuring that personnel remain fully attentive to the process being controlled. However, it is labor-intensive and impractical for tasks involving lengthy or tedious calculations. In contrast, automatic control is particularly suited for handling repetitive and time-consuming computations efficiently.

1.6 Analysis and synthesis problems

It is useful to introduce two important classes of problems that arise in automatic control, the discipline that studies how control operates and should be operated. They are the analysis problem and the synthesis problem:

- the analysis problem aims at understanding how a process operates;

- the synthesis problem aims at determining the controller that can be used to steer the process toward a desired behavior.

The analysis is a fundamental step, as we cannot design a controller without understanding how the process works. The analysis of a system makes possible to understand how the system operates either without any control (in open loop) or in the presence of a controller. It makes possible also to evaluate the performance of a system when a controller is applied. For the purpose of the analysis of a process, theoretical calculations as well as numerical simulations may be used.

The synthesis consists of the mathematical formulation of the control problem, the calculation of the controller mathematical model to use, and finally its physical realization.

Both analysis and synthesis are possible thanks to mathematical models, the central tool of automatic control.

1.7 Control technology

Nowadays all new controllers are realized in digital technology. The term microcontroller refers to an electronic digital device able to perform computation and to interface with the external world. The device is often equipped with analog-to-digital converters, which are devices able to transform an analog quantity into a digital signal, and with digital-to-analog converters, which are devices that make the digital signals produced by the microcontroller adapt to the external world, by transforming them into analog signals. In addition, typically microcontrollers also have the possibility to interface with external analog-to-digital and digital-to-analog converters as well as to a variety of other electronic devices.

However, in the past many other technologies were used to realize a controller. Indeed, the use of automatic control systems dates back to the dawn of human civilization, as exemplified by irrigation systems that distribute and regulate the flow of water to fields. However, the widespread adoption of automatic control systems arose from the necessity of managing vast amounts of energy. During the Industrial Revolution, the imperative to control steam engines provided a significant impetus for the development of automatic control systems. Well-known examples are the control system of T. Watt and the ones designed by J. Maxwell. The field of automatic control experienced further development in the years leading up to the Second World War. The control of aerial vehicles and communication systems was accompanied by methodological studies that greatly contributed to advancing the field.

Many scientists contributed to developing the theory of automatic control. Among those we mention Routh and Nyquist, as well as Lyapunov and Bellman. Some of the results that we will illustrate in these 20 lectures are due to these scientists.

Over the years from the Second World War to recent times, control systems have undergone a series of technological advancements for their implementation, including hydraulic, pneumatic, electronic (analog), and electronic (digital). As mentioned above, nowadays, digital technology is the most commonly used for implementing new automatic control systems.

2

Lecture 2—Mathematical models

In this lecture we will first introduce the general form to provide a mathematical description of a dynamical system and then the linear one. We will then derive the mathematical model of several mechanical, electrical, and electro-mechanical systems.

2.1   State equations of a system

The first task in control engineering is to understand the physical plant to be controlled. This understanding is achieved through the formulation of a mathematical model, which describes, often in an approximate way, the real behavior of the plant. The mathematical model allows the plant to be represented within a computer environment. It is derived from the physical laws governing the system or, alternatively, from real data obtained through an identification process.

A key aspect of automatic control is that the quantities that appear in the systems to control are not static, but evolve in time. As we mentioned in the first lecture, control is not possible without the analysis of the process. Therefore, in this lecture we will introduce the mathematical models of a system. These are dynamical models. In fact, to understand how a variable evolves in time, we need to describe the mathematical expression that determines its time derivative. For this reason, the models that we will consider in this lecture take the following form:

$$\dot{\mathbf{x}} = f(\mathbf{x}, \mathbf{u}, t)$$
$$\mathbf{y} = g(\mathbf{x}, \mathbf{u}, t) \tag{2.1}$$

where $\mathbf{x} \in \mathbb{R}^n$, $\mathbf{u} \in \mathbb{R}^m$, and $\mathbf{y} \in \mathbb{R}^p$. Here, we have considered that there are m inputs, all collected in the column vector $\mathbf{u}$. In an analogous way, there are p outputs that are collected in the column vector $\mathbf{y}$. Finally, to describe the system, we are using n variables, which are called *state variables* and are stacked into the column vector $\mathbf{x}$. The integer n is called the system order.

When $p > 1$ or $m > 1$, the system is called MIMO, which stands for multi-input-multi-output. When $p = 1$ and $m = 1$, the system is called SISO, which

DOI: 10.1201/9781003487289-2

stands for single-input-single-output. In this case the system is driven by a single input and provides a single output.

Eqs. (2.1) are very general, since $f : \mathbb{R}^n \times \mathbb{R}^m \times \mathbb{R} \to \mathbb{R}^n$ and $g : \mathbb{R}^n \times \mathbb{R}^m \times \mathbb{R} \to \mathbb{R}^p$ are two generic functions. Specifically, these functions cannot be totally arbitrary, as the system needs to provide a unique solution once fixed the system initial condition. Here, being interested to engineering problems, we give for granted the conditions for existence and uniqueness of the solution. In many cases, Eqs. (2.1) take a special form, which is simpler to analyze and for which many results have been obtained:

$$\begin{aligned} \dot{\mathbf{x}} &= \mathbf{A}\mathbf{x} + \mathbf{B}\mathbf{u} \\ \mathbf{y} &= \mathbf{C}\mathbf{x} + \mathbf{D}\mathbf{u} \end{aligned} \tag{2.2}$$

where $\mathbf{A} \in \mathbb{R}^{n\times n}$, $\mathbf{B} \in \mathbb{R}^{n\times m}$, $\mathbf{C} \in \mathbb{R}^{p\times n}$, and $\mathbf{D} \in \mathbb{R}^{p\times m}$ are matrices of constant coefficients.

Eqs. (2.2) describe a linear time-invariant (LTI) system. In the next sections of this lecture, we will focus on examples demonstrating how many systems can be described using Eqs. (2.2). Specifically, we will focus on mechanical, electrical, and electro-mechanical systems, but Eqs. (2.2) are applied in many different fields. In the next lectures, we will, instead, study the properties of these systems.

2.2 Models of mechanical systems

The Newton's law is key to derive the mathematical equations of a mechanical system. According to the Newton's law, we have that

$$\sum_i F_i = ma \tag{2.3}$$

where $\sum_i F_i$ represents the sum of all forces F_i acting on a body of mass m, and a its acceleration. If x indicates the position of the body in a fixed reference frame, then $a = \ddot{x}$. Eq. (2.3) applies in case of translations. In case of rotations, the Newton's law yields:

$$\sum_i \tau_i = I\dot{\omega} \tag{2.4}$$

where $\sum_i \tau_i$ is the sum of all torques τ_i acting on the body, and ω its angular speed, which in turn is $\omega = \dot{\theta}$, where θ is the angle of the body axis with respect to a fixed reference frame.

Example 2.1 _____

Model of cruise speed control.
As a first example, we consider the control of the cruise speed for a car (Fig. 2.1). Here we assume that the engine applies a force u to the car which has mass m. We assume that the car is subject to friction, which acts contrarily to motion and is proportional according to a factor b (the damping coefficient) to the car speed. The position of the car with respect to the fixed reference frame is indicated as x and illustrated along with all forces acting on the car body in the schematic diagram of Fig. 2.1.

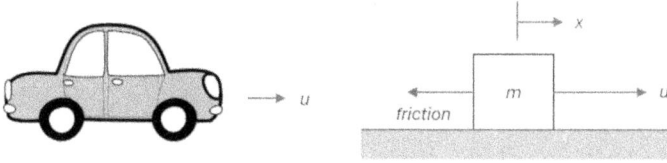

FIGURE 2.1
Cruise speed control in a car and scheme of the forces acting on the body of the car.

Applying Newton's law (2.3) to the car body, we get

$$u - b\dot{x} = m\ddot{x} \tag{2.5}$$

or equivalently

$$\ddot{x} + \frac{b}{m}\dot{x} = \frac{u}{m} \tag{2.6}$$

Since the focus is cruise speed control, the variable of interest is the speed $v = \dot{x}$. In terms of this variable, Eq. (2.6) is rewritten as follows:

$$\dot{v} + \frac{b}{m}v = \frac{u}{m} \tag{2.7}$$

or

$$\dot{v} = -\frac{b}{m}v + \frac{u}{m} \tag{2.8}$$

Let us now compare Eq. (2.6) with Eqs. (2.2), considering $y = v$. We notice that $n = 1$, $p = 1$, $m = 1$. Hence, the system is SISO and has order 1, so it is described by a single state variable, namely the car speed v. By comparison, we also conclude that the state matrices for this example are: $A = -\frac{b}{m}$, $B = \frac{1}{m}$, $C = 1$, and $D = 0$.

Example 2.2 _____

Quarter-car model.
We now discuss a simplified model of car suspensions. The model is known as the quarter-car model as it focuses on a single suspension and considers a quarter of the car mass, here indicated as m_2. The body associated with this mass is connected through a spring k_s and a bumper b to a second body, of mass m_1, representing the wheel. The spring and the bumper model the car suspension. Finally, the second body is connected to the terrain through a spring k_w, representing the elasticity of the tire. The terrain height is modeled as an input r to the system. The model does not need to account for the gravity force that would only shift the resting position of the springs.

The quarter-car model is illustrated in Fig. 2.2 along with the diagrams of the forces acting on the car and wheel bodies.

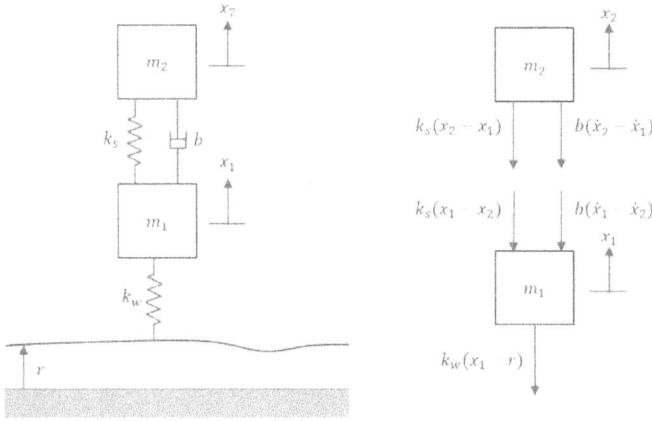

FIGURE 2.2
Quarter-car model and diagrams of the forces acting on the two bodies with masses m_1 and m_2.

Applying the Newton's law (2.3) to the two bodies of Fig. 2.2, we obtain the following equations:

$$b(\dot{x}_2 - \dot{x}_1) + k_s(x_2 - x_1) - k_w(x_1 - r) = m_1\ddot{x}_1$$
$$-k_s(x_2 - x_1) - b(\dot{x}_2 - \dot{x}_1) = m_2\ddot{x}_2 \tag{2.9}$$

We note that in these equations we find terms with the second derivatives of the position, while in the model (2.2) only first derivatives appear. To match Eqs. (2.2), let us now introduce two further variables: $x_3 = \dot{x}_1$ and $x_4 = \dot{x}_2$. Taking into account these definitions, we obtain

$$b(x_4 - x_3) + k_s(x_2 - x_1) - k_w(x_1 - r) = m_1\dot{x}_3$$
$$-k_s(x_2 - x_1) - b(x_4 - x_3) = m_2\dot{x}_4 \tag{2.10}$$

By reordering the terms, we get

$$\dot{x}_1 = x_3$$
$$\dot{x}_2 = x_4$$
$$\dot{x}_3 = -\frac{k_w+k_s}{m_1}x_1 + \frac{k_s}{m_1}x_2 - \frac{b}{m_1}x_3 + \frac{b}{m_1}x_4 + \frac{k_w}{m_1}r \tag{2.11}$$
$$\dot{x}_4 = \frac{k_s}{m_2}x_1 - \frac{k_s}{m_2}x_2 + \frac{b}{m_2}x_3 - \frac{b}{m_2}x_4$$

We can now compare Eq. (2.11) with Eqs. (2.2), considering as output the height of the car, namely $y = x_2$. The system we obtain is of order 4 ($n = 4$), with a single input ($u = r$, $m = 1$) and a single output ($y = x_2$, $p = 1$). The state vector is given by $\mathbf{x} = \begin{bmatrix} x_1 \\ x_2 \\ x_3 \\ x_4 \end{bmatrix}$. The state matrices are:

$$A = \begin{bmatrix} 0 & 0 & 1 & 0 \\ 0 & 0 & 0 & 1 \\ -\frac{k_w+k_s}{m_1} & \frac{k_s}{m_1} & -\frac{b}{m_1} & \frac{b}{m_1} \\ \frac{k_s}{m_2} & -\frac{k_s}{m_2} & \frac{b}{m_2} & -\frac{b}{m_2} \end{bmatrix} ; \quad B = \begin{bmatrix} 0 \\ 0 \\ \frac{k_w}{m_1} \\ 0 \end{bmatrix} ; \quad (2.12)$$

$$C = \begin{bmatrix} 0 & 1 & 0 & 0 \end{bmatrix} ; \qquad D = 0$$

Example 2.3

Spacecraft attitude control.

This example deals with a simplified model for attitude control of a satellite. Specifically, we consider the model for a single axis, as the other two axes of rotation can be dealt with in an analogous fashion.

The problem is schematically illustrated in Fig. 2.3. The rotation angle around an axis perpendicular to the book page is indicated as θ and measured with respect to an inertial frame of reference. This is the variable of interest of the problem, so, it is considered as the output of the system to control. The satellite is actuated by two thrusters, providing a torque equal to dF_c, where d is the distance of the thrusters from the center of the spacecraft. In addition, we also consider that the system is subject to a disturbance signal that applies a torque equal to M_d.

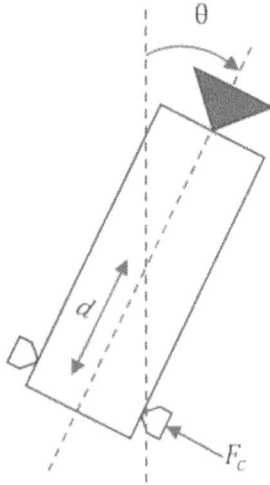

FIGURE 2.3

Attitude control of a satellite.

Considering the Newton's law for rotational motion of Eq. (2.4), we obtain

$$F_c d + M_d = I\ddot{\theta} \qquad (2.13)$$

This system has two inputs, $u_1 = F_c d$ and $u_2 = M_d$ ($m = 2$). The output is the angle $y = \theta$ ($p = 1$). Let us consider $x_1 = \theta$, and $x_2 = \dot{\theta}$. Then, we have

$$\begin{aligned} \dot{x}_1 &= x_2 \\ \dot{x}_2 &= \tfrac{1}{I}(u_1 + u_2) \end{aligned} \qquad (2.14)$$

The state matrices for this system read

$$A = \begin{bmatrix} 0 & 1 \\ 0 & 0 \end{bmatrix}; \quad B = \begin{bmatrix} 0 & 0 \\ \frac{1}{7} & \frac{1}{7} \end{bmatrix};$$
$$C = \begin{bmatrix} 1 & 0 \end{bmatrix}; \quad D = 0 \tag{2.15}$$

Example 2.4 _____

Pendulum.
Let us now consider the pendulum shown in Fig. 2.4. The pendulum has length l, mass m, and is actuated by a torque T_c. The pendulum angle is indicated as θ.

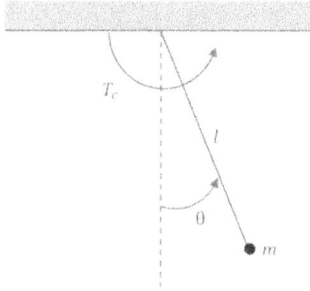

FIGURE 2.4
Schematic representation of a pendulum of length l and mass m.

Applying the Newton's law (2.4), we derive the following equation:

$$T_c - mgl \sin \theta = I\ddot{\theta} \tag{2.16}$$

that can be rewritten as

$$\ddot{\theta} + \frac{g}{l} \sin \theta = \frac{T_c}{ml^2} \tag{2.17}$$

We notice that this model is nonlinear because of the term $\sin\theta$. However, assuming small rotations, we can approximate $\sin\theta \approx \theta$, leading to the following expression:

$$\ddot{\theta} + \frac{g}{l}\theta = \frac{T_c}{ml^2} \tag{2.18}$$

Analogously to the previous example, we now define $x_1 = \theta$, and $x_2 = \dot{\theta}$. In addition, let us consider $u = T_c$ and $y = \theta$. Then, we have

$$\begin{aligned} \dot{x}_1 &= x_2 \\ \dot{x}_2 &= -\frac{g}{l}x_1 + \frac{1}{ml^2}u \end{aligned} \tag{2.19}$$

The state matrices for this system are

$$A = \begin{bmatrix} 0 & 1 \\ -\frac{g}{l} & 0 \end{bmatrix}; \quad B = \begin{bmatrix} 0 \\ \frac{1}{ml^2} \end{bmatrix};$$
$$C = \begin{bmatrix} 1 & 0 \end{bmatrix}; \quad D = 0 \tag{2.20}$$

Example 2.5 _____

Inverted pendulum.
Let us now consider the inverted pendulum that is illustrated in Fig. 2.5. Similarly to the previous example, the pendulum has length l, mass m, and is actuated by a torque T_c. The pendulum angle is indicated as θ.

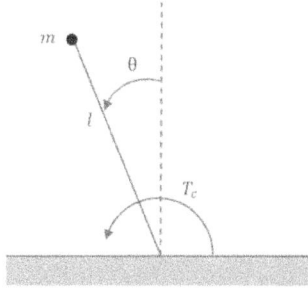

FIGURE 2.5
Schematic representation of an inverted pendulum of length l and mass m.

Applying the Newton's law (2.4) and rearranging the terms, we derive the following equation:

$$\ddot{\theta} - \frac{g}{l}\sin\theta = \frac{T_c}{ml^2} \tag{2.21}$$

while the linearized system reads

$$\ddot{\theta} - \frac{g}{l}\theta = \frac{T_c}{ml^2} \tag{2.22}$$

By letting $x_1 = \theta$, $x_2 = \dot{\theta}$, $u = T_c$ and $y = \theta$, we obtain the state equations of the linearized inverted pendulum

$$\begin{aligned} \dot{x}_1 &= x_2 \\ \dot{x}_2 &= \frac{g}{l}x_1 + \frac{1}{ml^2}u \end{aligned} \tag{2.23}$$

The state matrices for this system are

$$A = \begin{bmatrix} 0 & 1 \\ \frac{g}{l} & 0 \end{bmatrix}; \quad B = \begin{bmatrix} 0 \\ \frac{1}{ml^2} \end{bmatrix}; \\ C = \begin{bmatrix} 1 & 0 \end{bmatrix}; \quad D = 0 \tag{2.24}$$

2.3 Models of electrical systems

The state equations of electrical systems, composed of resistors, capacitors, and inductors, can be derived using Kirchhoff's laws. Kirchhoff's voltage law for a loop states that:

$$\sum_{j} v_j = 0 \tag{2.25}$$

where v_j for $j = 1, \ldots, q_V$ represents the voltage across the j-th component of the loop. Kirchhoff's current law for a node A of the circuit state that

$$\sum_{j} i_j = 0 \tag{2.26}$$

where i_j for $j = 1, \ldots, q_I$ is the current flowing through the j-th component connected to node A.

Next, we discuss two examples of models of electrical circuits.

Example 2.6 _____

RC circuit.
Consider the RC circuit illustrated in Fig. 2.6. We take the voltage across the capacitor as the state variable, which also correspond to the output of the circuit: $x = y = v_C$. Applying the Kirchhoff's voltage law to the loop formed by the voltage input, the resistor, and the capacitor, we obtain

$$v_{in} - Ri - v_C = 0 \tag{2.27}$$

Considering that $i = C \frac{dv_c}{dt}$ and letting $v_i n = u$, we get

$$\dot{x} = -\frac{1}{RC} x + \frac{1}{RC} u \tag{2.28}$$

The system order is one, so that A, B, C, and D are all scalar quantities:

$$A = -\frac{1}{RC}; B = \frac{1}{RC}; C = 1; D = 0 \tag{2.29}$$

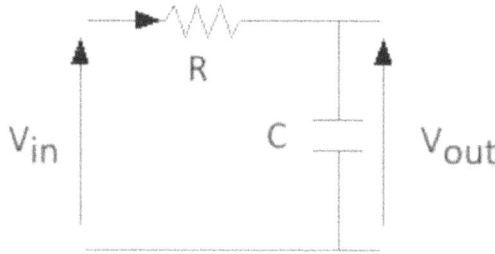

FIGURE 2.6
Electrical scheme of an RC circuit.

Example 2.7 _____

RLC circuit.

Consider the circuit illustrated in Fig. 2.7. The state variables are the current in the inductor and the voltage across the capacitor, namely $x_1 = i$ and $x_2 = v_c$. Applying the Kirchhoff's voltage law, we get that

$$v_{in} - Ri - v_L - v_C = 0 \qquad (2.30)$$

where v_L and v_C indicate the voltage across the inductor and the capacitor, respectively, and i the current. Taking into account that $v_L = L\frac{di}{dt}$, we obtain

$$\frac{di}{dt} = \frac{1}{L}(-Ri - v_C + v_{in}) \qquad (2.31)$$

Replacing $i = x_1$ and $u = v_{in}$, we get the dynamical equation of the first state variable:

$$\dot{x}_1 = -\frac{R}{L}x_1 - \frac{1}{L}x_2 + \frac{1}{L}v_{in} \qquad (2.32)$$

The dynamical equation of the other variable is obtained taking into account that $i = C\frac{dv_c}{dt}$. This yields

$$\dot{x}_2 = \frac{1}{C}x_1 \qquad (2.33)$$

Lastly, the output equation is

$$y = x_2 \qquad (2.34)$$

We conclude that the state matrices for the RLC circuit are

$$A = \begin{bmatrix} -\frac{R}{L} & -\frac{1}{L} \\ \frac{1}{C} & 0 \end{bmatrix}; \quad B = \begin{bmatrix} \frac{1}{L} \\ 0 \end{bmatrix}; \\ C = \begin{bmatrix} 0 & 1 \end{bmatrix}; \qquad D = 0 \qquad (2.35)$$

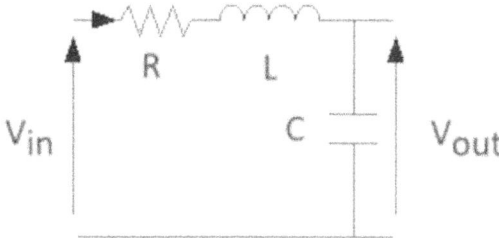

FIGURE 2.7
Electrical scheme of an RLC circuit.

2.4 Models of electro-mechanical systems

Electro-mechanical devices are systems that integrate both electrical and mechanical processes to perform their functions. Examples are generators, where electrical power is generated from a mechanical process, and motors, where, conversely, electrical power is used to produce a mechanical motion. In the next example, we discuss the model of a direct current (DC) motor.

Example 2.8 _____

Model of a DC motor.
Let us consider the DC motor schematically shown in Fig. 2.8. We assume that the rotor has a moment of inertia equal to J_m and its motion is described by the angle θ_m. While moving it induces a counter-electromotive force equal to $e = K_e\dot{\theta}_m$. The rotor is actuated by a coil L_a (with associated resistance R_a) where a voltage v_a is applied. This circuit generates a torque T that is equal to $T = K_t i_a$.

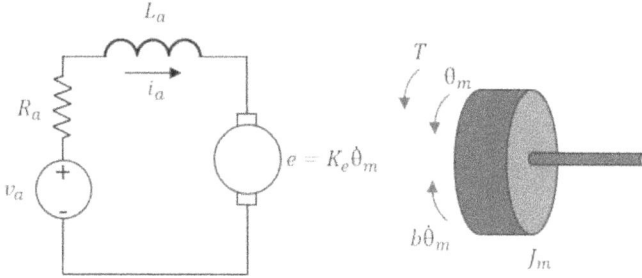

FIGURE 2.8
Illustrative scheme of a DC motor.

Taking into account Newton's law (2.4), for the DC motor we get

$$J_m\ddot{\theta}_m + b\dot{\theta}_m = K_t i_a \tag{2.36}$$

This equation has to be paired with the one describing the electrical circuit driving the rotor. This equation can be derived by applying the Kirchhoff's voltage law along with the constitutive equation for the inductor that relates the voltage across the component to the derivative of the current. In this way, one obtains

$$L_a\frac{di_a}{dt} + R_a i_a = v_a - k_e\dot{\theta}_m \tag{2.37}$$

For this system, the state vector will therefore include a current, an angle, and an angular speed. Specifically, the state vector is given by $\mathbf{x} = \begin{bmatrix} \theta_m \\ \omega_m \\ i_a \end{bmatrix}$ with $\omega_m = \dot{\theta}_m$.

Taking into account this definition, we can write

$$\begin{aligned} \dot{x}_1 &= x_2 \\ \dot{x}_2 &= -\frac{b}{J_m}x_2 + \frac{K_t}{J_m}x_3 \\ \dot{x}_3 &= -\frac{k_e}{L_a}x_2 - \frac{R_a}{L_a}x_3 + u/La \end{aligned} \tag{2.38}$$

This system is in the form of Eqs. (2.2) with the following state matrices:

$$A = \begin{bmatrix} 0 & 1 & 0 \\ 0 & -\frac{b}{J_m} & \frac{K_t}{J_m} \\ 0 & -\frac{k_e}{L_a} & -\frac{R_a}{L_a} \end{bmatrix} ; \quad B = \begin{bmatrix} 0 \\ 0 \\ \frac{1}{L_a} \end{bmatrix} ; \qquad (2.39)$$

$$C = \begin{bmatrix} 1 & 0 & 0 \end{bmatrix} ; \qquad D = 0$$

2.5 Exercises

1. Write the state matrices associated with the following system:

$$\begin{aligned} \dot{x}_1 &= x_1 + 3x_2 - x_3 \\ \dot{x}_2 &= x_2 + 4x_3 + 3u \\ \dot{x}_3 &= 3x_1 - 2x_2 - 2x_3 + 2u \end{aligned} \qquad (2.40)$$

with $y = x_1 - x_2$.

2. Write the state matrices associated with the following system:

$$\begin{aligned} \dot{x}_1 &= x_2 + x_4 + u_1 \\ \dot{x}_2 &= x_1 + x_3 + x_4 + u_2 \\ \dot{x}_3 &= -x_1 + 2x_2 - 5x_3 + x_4 \\ \dot{x}_4 &= x_1 - 2x_4 \end{aligned} \qquad (2.41)$$

with $y_1 = x_1$ and $y_2 = x_4$.

3. Consider the system in Fig. 2.1 and write a model where the state variables are the car position and velocity.

4. Consider the system in Fig. 2.9 and develop a model that accounts for frictional forces between the mass and the pavement (coefficient b_2).

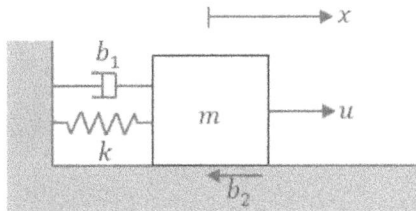

FIGURE 2.9
Mass-damper-spring system for Exercise 4.

5. Consider the system in Fig. 2.10 and develop a model considering the torque T as the input and the angle θ_2 as the output.

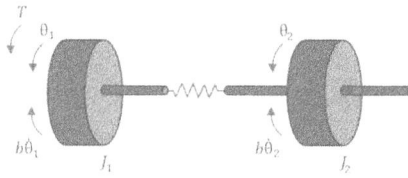

FIGURE 2.10
Mass-damper-spring system for Exercise 5.

3

Lecture 3—Brief introduction to MATLAB®

In this lecture, we illustrate some basic MATLAB® commands that are useful to define variables, indexing elements of vectors and matrices, and perform elementary operations. Lastly, we show how to define a linear time-invariant model in MATLAB®.

3.1 Vector and matrix definition

MATLAB® stands for MATrix LABoratory. We begin our discussion of MATLAB® with an example of how to define in MATLAB® a 3×3 matrix. To this aim, we use [and], to start and end the input of the coefficients. The coefficients in a single row can be separated either by a space or a comma ,, while a row is ended with a semicolon ;, as follows:

```
A=[1 2 4; 3 2 1; 3 1 2]
```

To avoid an output on the prompt, we can use the semicolon ; at the end of the command.

```
A=[1 2 4; 3 2 1; 3 1 2];
```

This is useful when the size of the vector or the matrix considered is large. As an example, the reader can check the different result obtained with

```
t=[0:0.1:1000]
```

or

```
t=[0:0.1:1000];
```

The two commands define a vector (representing time) whose elements start from 0 up to 1000 with steps equal to 0.1. In the first case, a large number of values are printed in the prompt, while in the second case the variable is created but no output in the prompt is produced.

DOI: 10.1201/9781003487289-3

3.2 Commands to define specific matrices: ones, zeros, eye, rand, randn

We now illustrate the MATLAB® commands to define a few specific matrices. Such commands are: `ones`, `zeros`, `eye`, `rand`, `randn`.

We start with the following example

```
A2=ones(2,3)
```

that creates a 2×3 matrix filled with all elements equal to 1. Eventually, rather than specifying the number of rows (the first input of the command) and that of columns (the second input of the command), for square matrices only a single input can be provided. For example, the command

```
A3=ones(5)
```

creates a 5×5 matrix having all elements equal to 1.

The syntax of the other commands is similar. For example, the following commands

```
A4=zeros(2,7)
A5=eye(3)
A6=rand(3)
A7=randn(3)
```

create a 2×7 matrix A_4 with all zero elements, a 3×3 matrix A_5 that is the identity matrix of order 3, a 3×3 matrix A_6 with random values drawn from a uniform distribution in $[0, 1]$, and a 3×3 matrix A_7 with random values drawn from a normal distribution with zero mean and unit standard deviation. So, the commands `rand` and `randn` both produce matrices with random values, but differ for the type of distribution from which the data are drawn.

An instructive exercise is to check the different type of distribution by looking at the hystogram of a large number of elements generated with `rand` and `randn`. This can be done with the following commands:

```
x1=rand(1000000,1);
figure,hist(x1,100)
x2=randn(10000000,1);
figure,hist(x2,100)
```

Here, we note that the second input parameter in the command `hist` specifies the number of bins used in the plot.

MATLAB® exercise 3.1 _____

Consider the following matrix

$$
A = \begin{bmatrix}
2 & 2 & 2 & 2 & 2 \\
2 & 3 & 3 & 3 & 3 \\
2 & 3 & 3 & 3 & 3 \\
2 & 3 & 3 & 3 & 3
\end{bmatrix}
\tag{3.1}
$$

and define it in MATLAB® in a compact way.

To this aim, we use the commands described above. In fact, we can use three blocks, each created with `ones`, as follows:

```
A=[2*ones(1,5); 2*ones(3,1) 3*ones(3,4)]
```

3.3 Indexing the elements of a matrix

We now discuss how one or more coefficients of a matrix can be indexed. The following commands illustrate how to index the coefficient A_{23} of matrix A

```
A(2,3)
```

In a similar way, it is possible to access to a mini-block of a matrix

```
A([1 2],2)
```

The command returns the elements A_{12} and A_{22}.
As another example, the following command:

```
A(1:1:3,1:1:3)
```

extracts the mini-block of A containing elements having row indices from 1 to 3, and column indices from 1 to 3. The same result can be obtained as follows:

```
A(1:3,1:3)
```

In fact, using the command : if the step size is not specified, a unit step size is assumed. When the starting and ending points are not specified, then all elements are indexed. For instance, the following commands:

```
A=rand(6)
A(:,2)
```

define a 6×6 matrix of random values and then extract the second column.
Finally, using

```
A(:)
```

returns all elements of the matrix A taken columnwise.

3.4 Basic operations between matrices and element-by-element operations

Basic operations between matrices can be done straightforwardly. For instance, the following commands

```
A1=rand(2)
A2=rand(2)
A1+A2
```

define two 2×2 matrices of random elements, and then sum them. To perform the row-column product, the following command can be used

```
A1*A2
```

Instead, to multiply element-by-element the coefficients of the two matrices the command to use is

```
A1.*A2
```

Placing a dot . before the operator instructs MATLAB® to perform the operation in a element-by-element fashion. For instance, the command

```
A1^2
```

produces the square of the matrix A_1, whereas

```
A1.^2
```

generates a matrix where the generic ij element is the square of the corresponding ij element of the matrix A_1.

3.5   Plot of a function of a variable

We now illustrate how to plot a function of a variable. As an example we consider $y = \ln x$:

```
x=logspace(-4,5,1000);
y=log(x);
figure,plot(x,y)
xlabel('x')
ylabel('y')
```

The command `y=logspace(-4,5,1000);` generates a vector of 1000 logarithmically equally spaced points, starting from 10^{-4} up to 10^5. Then, the variable y is computed, and finally the two vectors x and y are plotted one as a function of the other.

As a second example, we consider the function $y = x^3$:

```
x=[-10:0.1:10];
y=x.^3;
figure,plot(x,y)
xlabel('x')
ylabel('y')
```

We note that this time, the vector x is generated using linearly equally spaced points.

3.6   Definition of an LTI system in MATLAB

In this section, we illustrate how to define a LTI system in MATLAB®, starting from its state matrices. To this aim, we consider the DC motor model studied in Sec. 2.8. The state matrices of the system are given in Eq. (2.39).

First, we define the system parameters (here we use exemplified, normalized values):

```
b=0.5;
Jm=2;
La=3
Ra=0.8;
Kt=2.7;
ke=1;
```

Next, we define the state matrices

```
A=[0  1  0;  0  -b/Jm  Kt/Jm;  0  -ke/La  -Ra/La]
B=[0;  0;  1/La]
C=[1  0  0]
D=0
```

Once given the state matrices, the LTI system is defined using the command **ss**, which stands for state-space (representation):

```
DCmotormodel=ss(A,B,C,D)
```

This command creates an object of LTI type that can be used in several functions taking as an input an LTI object. The object stores the state matrices that can be retrieved using the following commands:

```
DCmotormodel.A
DCmotormodel.B
DCmotormodel.C
DCmotormodel.D
```

4

Lecture 4—Laplace transform

In this lecture, we introduce the Laplace transform that is an important mathematical tool for the study of systems described by continuous-time linear models. We start from the definition and a few properties and then focus on its applications to LTI systems. We end the lecture with a MATLAB® example.

4.1 The Laplace transform of a signal

Given a signal $x(t)$ defined such that $x(t) = 0 \; \forall t < 0$, its Laplace transform is given by

$$X(s) = \int_0^\infty x(t)e^{-st}dt \qquad (4.1)$$

where s is a complex variable, i.e., $s = \sigma + j\omega$. The interested reader may refer to books on mathematics for a rigorous and detailed treatment of the Laplace transform. Here, we focus on the signal having finite energy for which we can assume that the Laplace transform always exists.

An important aspect of the Laplace transform is that it enables the transition from the time domain, where the independent variable is real (time t), to the Laplace domain, where the independent variable is complex (represented by $s = \sigma + j\omega$). As we will see in the next lectures, certain operations become particularly straightforward in the Laplace domain.

We will also restrict the discussion on the properties of the Laplace transform to the ones more useful in the following, without any pretense of completeness.

- The Laplace transform is a *linear* operator. This means that, given two real numbers a_1 and a_2, if $x(t) = a_1 x_1(t) + a_2 x_2(t)$, then

$$X(s) = a_1 X_1(s) + a_2 X_2(s)$$

DOI: 10.1201/9781003487289-4

- The Laplace transform of a time-delayed signal, that is, $x(t) = x_1(t - \tau)$, where $\tau > 0$ and $x_1(t) = 0 \; \forall t < 0$, is given by

$$X(s) = e^{-s\tau} X_1(s)$$

- The Laplace transform of the signal $x(t) = e^{at} x_1(t)$ with $a \in \mathbb{R}$, is

$$X(s) = X_1(s - a)$$

- The Laplace transform of the derivative of a signal, $x(t) = \frac{dx_1(t)}{dt}$ is given by

$$X(s) = sX_1(s) - x_1(0)$$

- For the derivative of order n, namely for the signal $x(t) = \frac{d^{(n)} x_1(t)}{dt^n}$, one has that

$$X(s) = s^n X_1(s) - s^{n-1} x_1(0) - \ldots - \frac{d^{n-1} x_1(t)}{dt^{n-1}} \bigg|_{t=0}$$

- The Laplace transform of the integral of a function, that is, $x(t) = \int\limits_0^t x_1(\tau) d\tau$ is

$$X(s) = \frac{1}{s} X_1(s)$$

- The Laplace transform of the convolution between two signals, namely $x(t) = \int\limits_0^\infty x_1(t - \tau) x_2(\tau) d\tau$ is

$$X(s) = X_1(s) X_2(s)$$

Next, we illustrate the Laplace transform of several important signals. These can be obtained by direct application of the definition of the Laplace transform or straightforward use of the properties discussed above. Here, we indicate the Laplace transform as follows: $\mathcal{L}(x(t)) = X(s)$.

- $\mathcal{L}(e^{at}) = \frac{1}{s-a}$;

- $\mathcal{L}(\text{step}(t)) = \frac{1}{s}$, where $\text{step}(t) = 1$ for $t > 0$, and $\text{step}(t) = 0$ for $t \leq 0$;

- $\mathcal{L}(t^n) = \frac{n!}{s^{n+1}}$;

- $\mathcal{L}(t^n e^{at}) = \frac{n!}{(s-a)^{n+1}}$;

- $\mathcal{L}(\cos \omega t) = \frac{s}{s^2 + \omega^2}$;

- $\mathcal{L}(\sin \omega t) = \frac{\omega}{s^2 + \omega^2}$;

- $\mathcal{L}(e^{at} \cos \omega t) = \frac{s-a}{(s-a)^2 + \omega^2}$;

- $\mathcal{L}(e^{at} \sin \omega t) = \frac{\omega}{(s-a)^2 + \omega^2}$.

Finally, for the Dirac delta function (the unit impulse), we have the result that

$$\mathcal{L}(\delta(t)) = 1$$

4.2 Examples of Laplace transform

Here we consider a few examples of calculation of Laplace transforms.

Example 4.1 _____

Calculate the Laplace transform of $x(t) = 4t^2 - 3\cos 2t + 5e^{-t}$.
We have that

$$\mathcal{L}(4t^2 - 3\cos 2t + 5e^{-t}) = 4\mathcal{L}(t^2) - 3\mathcal{L}(\cos 2t) + 5\mathcal{L}(e^{-t})$$
$$= \frac{8}{s^3} - \frac{3s}{s^2+4} + \frac{5}{s+1}$$

Example 4.2 _____

Calculate the Laplace transform of $x(t) = e^{-t}\cos 3t$.
We have that

$$\mathcal{L}(e^{-t}\cos 3t) = \frac{s+1}{(s+1)^2+9}$$

Example 4.3 _____

Calculate the Laplace transform of $x(t) = (t-2)^3\text{step}(t-2))$.
We have that

$$\mathcal{L}((t-2)^3\text{step}(t-2)) = \frac{6}{s^4}e^{-2s}$$

Example 4.4 _____

Calculate the Laplace transform of $x(t) = \int_0^t \sin(2\tau)d\tau$.

We have that

$$\mathcal{L}(\int_0^t \sin(2\tau)d\tau)) = \frac{2}{s(s^2+4)}$$

Example 4.5 _____

Calculate the Laplace transform of $x(t) = 5t - 3$.
We have that

$$\mathcal{L}(5t-3) = \frac{2}{s^2} - \frac{3}{s}$$

Example 4.6 _____

Consider the signal $x(t)$ shown in Fig. 4.1. It represents a finite time pulse signal. Since it can be expressed as $x(t) = \text{step}(t) - \text{step}(t-T)$, its Laplace transform is given by

$$\mathcal{L}(x(t)) = \frac{1}{s}\left(1 - e^{-sT}\right)$$

Example 4.6 explains why the Laplace transform of the Dirac delta function is unit. In fact, if we consider a finite impulse of duration $T = \varepsilon$ and amplitude $1/\varepsilon$, that is, $\delta_\varepsilon(t) = \frac{1}{\varepsilon}$ for $0 < t \leq \varepsilon$, and $\delta_\varepsilon(t) = 0$ otherwise, using the result of Example 4.6, we have that

$$\mathcal{L}(\delta_\varepsilon(t)) = \frac{1}{s\varepsilon}(1 - e^{-s\varepsilon}).$$

Considering the behavior for $\varepsilon \to 0$ we find that $\mathcal{L}(\delta(t)) = 1$.

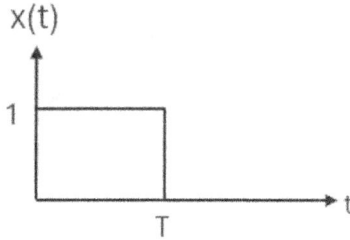

FIGURE 4.1
Finite time pulse signal.

Example 4.7 _____

Consider now the signal $x(t)$ shown in Fig. 4.2, that represents a triangular pulse signal. Since it can be expressed as $x(t) = \text{ramp}(t) - 2\text{ramp}(t-1) + \text{ramp}(t-2)$, its Laplace transform is

$$\mathcal{L}(x(t)) = \tfrac{1}{s^2}\left(1 - 2e^{-s} + e^{-2s}\right)$$

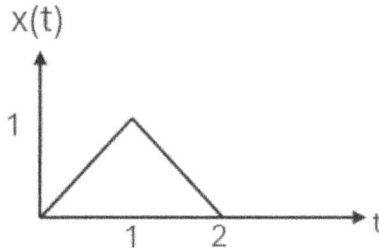

FIGURE 4.2
Triangular pulse signal.

4.3 Step response of the model for cruise speed control

Here, we consider again the model for cruise speed control, described by (see Sec. 2.1):

$$\dot{v} = -\frac{b}{m}v + \frac{u}{m} \tag{4.2}$$

and apply the Laplace transform to calculate the time evolution of the velocity v, when the input is a step function, namely $u(t) = U\text{step}(t)$, that is, when the motor is switched on and a constant unit force of amplitude U is applied to the car body. We suppose that at $t = 0$ the car is not moving, that is, $v(0) = 0$.

By applying the Laplace transform to the left and right side of (4.2), we get

$$\mathcal{L}(\dot{v}) = \mathcal{L}\left(-\frac{b}{m}v + \frac{u}{m}\right) \Rightarrow \tag{4.3}$$

$$sV(s) - v(0) = -\frac{b}{m}V(s) + \frac{1}{m}\frac{U}{s} \Rightarrow \tag{4.4}$$

$$\left(s + \frac{b}{m}\right)V(s) = \frac{1}{m}\frac{U}{s} \Rightarrow \tag{4.5}$$

$$V(s) = \frac{U/m}{s\left(s + \frac{b}{m}\right)} \tag{4.6}$$

Now we write $V(s)$ as the sum of two fraction, whose numerators R_1 and R_2 have to be calculated:

$$V(s) = \frac{R_1}{s} + \frac{R_2}{s + \frac{b}{m}} \Rightarrow \tag{4.7}$$

$$V(s) = \frac{R_1\left(s + \frac{b}{m}\right) + R_2 s}{s\left(s + \frac{b}{m}\right)} \tag{4.8}$$

The numerator of (4.6) and the one of (4.8) should be equal

$$\frac{U}{m} = R_1\left(s + \frac{b}{m}\right) + R_2 s \tag{4.9}$$

Polynomial identity requires that all coefficients of all powers of the two polynomials are equal. Therefore, R_1 and R_2 can be found solving the following equations:

$$\begin{cases} \frac{U}{m} &= R_1\frac{b}{m} \\ 0 &= R_1 + R_2 \end{cases} \tag{4.10}$$

which give

$$\begin{cases} R_1 &= \frac{U}{b} \\ R_2 &= -\frac{U}{b} \end{cases} \tag{4.11}$$

Replacing these values in (4.7), we obtain

$$V(s) = \frac{U}{b}\frac{1}{s} - \frac{U}{b}\frac{1}{s + \frac{b}{m}} \tag{4.12}$$

By taking into account that the Laplace transform is a linear operator and recalling that $\mathcal{L}(\text{step}(t)) = \frac{1}{s}$ and $\mathcal{L}(e^{at}) = \frac{1}{s-a}$, we get the evolution in time of the state variable:

$$v(t) = \frac{U}{b}\left(\text{step}(t) - e^{-\frac{b}{m}t}\right) \tag{4.13}$$

4.4 Inverse Laplace transform of rational functions

The inverse Laplace transform is the operation that, given the Laplace transform of a signal, returns the original signal. For example, given $X(s) = \frac{1}{s}$, then $x(t) = \mathcal{L}^{-1}\left(\frac{1}{s}\right) = \text{step}(t)$.

In automatic control, it is particularly important to learn how to calculate the inverse Laplace transform of rational functions, namely $X(s)$ that can be expressed as the ratio of two polynomials in s:

$$X(s) = \frac{N_x(s)}{D_x(s)} \tag{4.14}$$

where $D_x(s) = \prod_{i=1}^{r}(s - \lambda_i)^{k_i}$. Here, λ_i for $i = 1,\ldots,r$ represent the roots of the polynomial in the denominator, and k_i the multiplicity of each of these roots. Let n_x be the order of the polynomial, then $n_x = \sum_{i=1}^{r} k_i$.

The first step is to expand $X(s)$ in the sum of simple fractions

$$X(s) = \sum_{i=1}^{r}\left[\frac{R_{i,1}}{(s - \lambda_i)} + \frac{R_{i,2}}{(s - \lambda_i)^2} + \cdots + \frac{R_{i,k_i}}{(s - \lambda_i)^{k_i}}\right] \tag{4.15}$$

The constants $R_{i,j}$ are called the residues, and can be calculated as follows. For $j = k_i$, we have that

$$R_{i,k_i} = \lim_{s \to \lambda_i}(s - \lambda_i)^{k_i}X(s) \tag{4.16}$$

and for $j = 1,\ldots,k_i - 1$, we have that

$$R_{i,j} = \frac{1}{(k_i - j)!}\lim_{s \to \lambda_i}\frac{d^{(k_i-j)}}{ds^{(k_i-j)}}\left[(s - \lambda_i)^{k_i}X(s)\right] \tag{4.17}$$

At this point, taking into account that

$$\mathcal{L}^{-1}\left(\frac{R_{i,2}}{(s-\lambda_i)}\right) = R_{i,1}e^{\lambda_i t}$$
$$\mathcal{L}^{-1}\left(\frac{R_{i,2}}{(s-\lambda_i)^2}\right) = R_{i,2}te^{\lambda_i t}$$
$$\cdots$$
$$\mathcal{L}^{-1}\left(\frac{R_{i,k_i}}{(s-\lambda_i)^{k_i}}\right) = \frac{R_{i,k_i}}{(k_i-1)!}t^{k_i-1}e^{\lambda_i t} \tag{4.18}$$

we have that

$$\mathcal{L}^{-1}(X(s)) = \sum_{i=1}^{r} \left[R_{i,1}e^{\lambda_i t} + R_{i,2}te^{\lambda_i t} + \ldots + R_{k_i-1}t^{k_i-1}e^{\lambda_i t} \right] \tag{4.19}$$

Example 4.8 _____

Calculate the inverse Laplace transform of $X(s) = \frac{s^3+3s^2+s+3}{s^2(s^2+3s+2)}$.

First, we rewrite $X(s)$ factorizing the denominator

$$X(s) = \frac{s^3 + 3s^2 + s + 3}{s^2(s+1)(s+2)} \tag{4.20}$$

Then, we expand $X(s)$ in simple fractions

$$X(s) = \frac{R_{1,1}}{s} + \frac{R_{1,2}}{s^2} + \frac{R_2}{s+1} + \frac{R_3}{s+2} \tag{4.21}$$

We calculate the residues as follows:

$$R_{1,2} = \lim_{s \to 0} s^2 X(s) = \lim_{s \to 0} \frac{s^3 + 3s^2 + s + 3}{(s+1)(s+2)} = \frac{3}{2} \tag{4.22}$$

$$R_{1,1} = \lim_{s \to 0} \frac{d}{ds} s^2 X(s) = \lim_{s \to 0} \frac{d}{ds} \left(\frac{s^3 + 3s^2 + s + 3}{(s+1)(s+2)} \right) = -\frac{7}{4} \tag{4.23}$$

$$R_2 = \lim_{s \to -1} (s+1)X(s) = \lim_{s \to 0} s^2 \frac{s^3 + 3s^2 + s + 3}{s^2(s+2)} = 4 \tag{4.24}$$

$$R_3 = \lim_{s \to -2} (s+2)X(s) = \lim_{s \to 0} s^2 \frac{s^3 + 3s^2 + s + 3}{s^2(s+1)} = -\frac{5}{4} \tag{4.25}$$

Replacing these values in the expansion in simple fractions of $X(s)$, we get

$$X(s) = -\frac{7}{4}\frac{1}{s} + \frac{3}{2}s^2 + \frac{4}{s+1} - \frac{5}{4}\frac{1}{s+2} \tag{4.26}$$

and so

$$x(t) = -\frac{7}{4}\text{step}(t) + \frac{3}{2}t + 4e^{-t} - \frac{5}{4}e^{-2t} \tag{4.27}$$

In case of roots that are complex and conjugates (complex roots necessarily appear in pairs of complex and conjugate numbers, as the coefficients of $D_x(s)$ are real), to avoid calculations involving residues that are complex, it is possible to use a different expansion of $X(s)$. In this case, complex roots are not factorized, but the second-order polynomial associated with them, namely $(s - \lambda_i)(s - \bar{\lambda}_i) = (s - a)^2 + b^2$ with $\lambda_i = a + jb$. By doing so, $X(s)$ is expanded as:

$$X(s) = \sum_{i=1}^{r_r} \left[\frac{R_{i,1}}{(s-\lambda_i)} + \frac{R_{i,2}}{(s-\lambda_i)^2} + \cdots + \frac{R_{i,k_i}}{(s-\lambda_i)^{k_i}} \right] + \\ \sum_{i=1}^{r_c} \left[\frac{A_{i,1}s+B_{i,1}}{(s-a)^2+b^2} + \frac{A_{i,2}s+B_{i,2}}{((s-a)^2+b^2)^2} + \cdots + \frac{A_{i,k_i}s+B_{i,k_i}}{((s-a)^2+b^2)^{k_i}} \right] \tag{4.28}$$

where r_r indicates the number of real roots, and r_c the number of pairs of complex and conjugate roots. The residues $R_{i,j}$ are calculated using Eqs. (4.16) and (4.17), while the coefficients $A_{i,j}$ and $B_{i,j}$ by equating the polynomials in the numerator of $X(s)$ and of its expansion. Lastly, the inverse Laplace transform of $X(s)$ is calculated taking into account that

$$\mathcal{L}^{-1}\left(\frac{b}{(s-a)^2+b^2}\right) = e^{at}\sin(bt)$$

$$\mathcal{L}^{-1}\left(\frac{s-a}{(s-a)^2+b^2}\right) = e^{at}\cos(bt)$$

$$\mathcal{L}^{-1}\left(\frac{b}{((s-a)^2+b^2)^2}\right) = te^{at}\sin(bt)$$

$$\mathcal{L}^{-1}\left(\frac{s-a}{((s-a)^2+b^2)^2}\right) = te^{at}\cos(bt) \qquad (4.29)$$

$$\cdots$$

$$\mathcal{L}^{-1}\left(\frac{b}{((s-a)^2+b^2)^{k_i}}\right) = \frac{t^{k_i-1}}{(k_i-1)!}e^{at}\sin(bt)$$

$$\mathcal{L}^{-1}\left(\frac{s-a}{((s-a)^2+b^2)^{k_i}}\right) = \frac{t^{k_i-1}}{(k_i-1)!}e^{at}\cos(bt)$$

Example 4.9

Calculate the inverse Laplace transfrom of $X(s) = \frac{s+2}{s(s^2+s+1)}$.

The roots of the denominator of $X(s)$ are $\lambda_1 = 0$ and $\lambda_{2,3} = -\frac{1}{2} \pm j\frac{\sqrt{3}}{2}$. Therefore, we expand $X(s)$ as follows:

$$X(s) = \frac{R_1}{s} + \frac{As+B}{s^2+s+1} \qquad (4.30)$$

We calculate R_1 using Eq. (4.16)

$$R_1 = \lim_{s\to 0} sX(s) = \lim_{s\to 0} \frac{s+2}{s^2+s+1} = 2 \qquad (4.31)$$

Replacing $R_1 = 2$ in the expression of $X(s)$, we get

$$X(s) = \frac{2}{s} + \frac{As+B}{s^2+s+1} = \frac{2s^2+2s+2+As^2+Bs}{s(s^2+s+1)} \qquad (4.32)$$

Equating the coefficients of the numerators of $X(s) = \frac{s+2}{s(s^2+s+1)}$ and $X(s) = \frac{2s^2+2s+2+As^2+Bs}{s(s^2+s+1)}$, we have that

$$\begin{aligned} A+2 &= 0 \\ 2+B &= 1 \\ 2 &= 2 \end{aligned} \qquad (4.33)$$

that yields: $A = -2$ and $B = -1$. Replacing these values in the expression of $X(s)$ we have that

$$X(s) = \frac{2}{s} - \frac{2s+1}{s^2+s+1} = \frac{2}{s} - 2\frac{s+\frac{1}{2}}{\left(s+\frac{1}{2}\right)^2+\frac{3}{4}} \qquad (4.34)$$

This yields that

$$x(t) = 2\text{step}(t) - 2e^{-\frac{t}{2}}\cos\left(\frac{\sqrt{3}}{2}t\right) \qquad (4.35)$$

4.5   Calculation of the step response of the model for cruise speed control in MATLAB

MATLAB® offers several functions to analyze and simulate dynamical systems, including how to calculate the step response for systems defined through the command `ss` that we have introduced in the previous lecture. These functions will be discussed in detail in the next lectures. However, already at this point, it is instructive to discuss a method to obtain the step response of a given system. As an example, we will refer to the model of cruise speed control.

Let us consider the system described by Eq. (4.2) with $U = 2$, $b = 0.5$, and $m = 10$.

In the following, we calculate the step response by performing the inverse Laplace transform of Eq. (4.6) using the MATLAB® command `ilaplace`.

First, we define the numerical parameters appearing in the system dynamics.

```
U=2;
b=0.5;
m=10;
```

Then, we define the symbolic variables used in the calculations, namely the complex variable s and time t.

```
syms s t
```

Lastly, we calculate the inverse Laplace transform

```
ilaplace(U/m/(s*(s+b/m)))
```

The result matches Eq. (4.13) with the parameters taking the above mentioned values.

Notice that, to calculate the step response with the symbolic parameters U, b, and m, the following commands can be used

```
syms s t U m b
ilaplace(U/m/(s*(s+b/m)))
```

4.6   Exercises

1. Calculate the Laplace transform of the signal $u(t) = \text{step}(t-2) + e^{t-2}\text{step}(t-2) + 2e^{-3t}\sin(4t)$.

2. Calculate the Laplace transform of the signal $u(t) = e^t \sin(3t) + 2e^t \cos(3t)$.

3. Calculate the Laplace transform of the signal $u(t) = \frac{d}{dt}\left(e^{-5t} + 4\cos t\right)$.

4. Calculate the inverse Laplace transform of $X(s) = \frac{s+3}{(s+1)(s^2+3s+3)}$.

5. Calculate the inverse Laplace transform of $X(s) = \frac{s}{(s+2)(s+3)(s+5)}$.

5

Lecture 5—Continuous-time linear systems

In this lecture, we examine linear time-invariant systems that are continuous in time. We introduce their representation in the time domain and illustrate how to calculate the evolution of the key variables of these systems and its relevant properties.

5.1 State-space representation

Linear time-invariant (LTI) systems with continuous time, or shortly *continuous-time linear systems*, are described by the following state equations:

$$\dot{\mathbf{x}} = A\mathbf{x} + B\mathbf{u}$$
$$\mathbf{y} = C\mathbf{x} + D\mathbf{u} \tag{5.1}$$

where $A \in \mathbb{R}^{n \times n}$, $B \in \mathbb{R}^{n \times m}$, $C \in \mathbb{R}^{p \times n}$, and $D \in \mathbb{R}^{p \times m}$ are matrices of constant coefficients. Eq. (5.1) are known as the state-space form or state-space representation of the system, while the matrices A, B, C, and D as the state matrices. Here, all quantities—the state variables, the input, and the output—depend on time $t \in \mathbb{R}$, that is, $\mathbf{x} = \mathbf{x}(t)$, $\mathbf{u} = \mathbf{u}(t)$, and $\mathbf{y} = \mathbf{y}(t)$. The state-space representation, therefore, provides the mathematical model of a continuous-time linear system in the time domain.

Systems that can be written as in Eqs. (5.1) are called proper systems. When $D = 0$ they are said strictly proper systems.

The state-space representation of a system is not unique. In describing the same system, in fact, different state variables may be selected. An example is if we use different scaling factors or sign for the state variables. More in general, we can move from one representation to another where the state variables are a linear combination of the previous ones. Suppose then to consider a new set of state variables $\tilde{\mathbf{x}}$ defined through the relationship:

$$\tilde{\mathbf{x}} = T^{-1}\mathbf{x} \tag{5.2}$$

where T is an invertible $n \times n$ matrix.

DOI: 10.1201/9781003487289-5

Since $\dot{\tilde{\mathbf{x}}} = T^{-1}\dot{\mathbf{x}}$, we have that

$$
\begin{aligned}
\dot{\tilde{\mathbf{x}}} &= T^{-1}(A\mathbf{x} + B\mathbf{u}) \\
\mathbf{y} &= C\mathbf{x} + D\mathbf{u}
\end{aligned}
\tag{5.3}
$$

On the other hand, since the relationship (5.2) is invertible, we have that

$$
\mathbf{x} = T\tilde{\mathbf{x}}
\tag{5.4}
$$

and hence:

$$
\begin{aligned}
\dot{\tilde{\mathbf{x}}} &= T^{-1}AT\tilde{\mathbf{x}} + T^{-1}B\mathbf{u} \\
\mathbf{y} &= CT\tilde{\mathbf{x}} + D\mathbf{u}
\end{aligned}
\tag{5.5}
$$

Summing up, we have obtained an equivalent representation in state-space form:

$$
\begin{aligned}
\dot{\tilde{\mathbf{x}}} &= \tilde{A}\tilde{\mathbf{x}} + \tilde{B}\mathbf{u} \\
\mathbf{y} &= \tilde{C}\tilde{\mathbf{x}} + \tilde{D}\mathbf{u}
\end{aligned}
\tag{5.6}
$$

where

$$
\begin{aligned}
\tilde{A} &= T^{-1}AT \\
\tilde{B} &= T^{-1}B \\
\tilde{C} &= CT \\
\tilde{D} &= D
\end{aligned}
\tag{5.7}
$$

It is important to note that the change of variables concerns the internal variables used to describe the system, but clearly not the input and output of the system that remain unvaried.

Another important property is that A and $\tilde{A}$ are linked by a similarity relationship, namely $\tilde{A} = T^{-1}AT$. Therefore, they have the same eigenvalues.

The eigenvalues of the state matrix A are invariant to state transformation. They are called the *system eigenvalues*.

Example 5.1 _____

Consider the mass-spring system composed by two bodies connected by a spring, as in Fig. 5.1.

By applying the Newton's law, we derive the system equations

$$
\begin{aligned}
\dot{x}_1 &= x_3 \\
\dot{x}_2 &= x_4 \\
\dot{x}_3 &= -\frac{k}{m_1}x_1 + \frac{k}{m_1}x_2 + \frac{1}{m_1}u_1 \\
\dot{x}_4 &= \frac{k}{m_2}x_1 - \frac{k}{m_2}x_2 + \frac{1}{m_2}u_2
\end{aligned}
\tag{5.8}
$$

with $y = x_1$.

The state matrices associated with this state-space representation are

$$
A = \begin{bmatrix} 0 & 0 & 1 & 0 \\ 0 & 0 & 0 & 1 \\ -\frac{k}{m_1} & \frac{k}{m_1} & 0 & 0 \\ \frac{k}{m_2} & -\frac{k}{m_2} & 0 & 0 \end{bmatrix}; \quad B = \begin{bmatrix} 0 & 0 \\ 0 & 0 \\ \frac{1}{m_1} & 0 \\ 0 & \frac{1}{m_2} \end{bmatrix};
$$
$$
C = \begin{bmatrix} 1 & 0 & 0 & 0 \end{bmatrix}; \quad D = 0
\tag{5.9}
$$

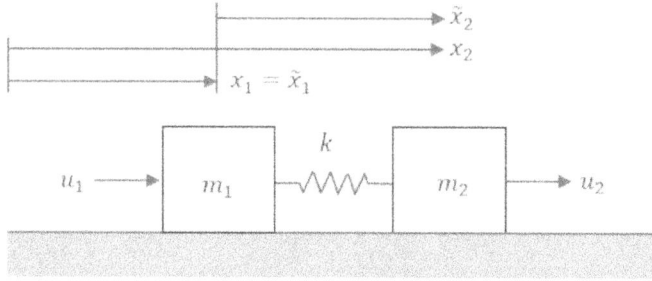

FIGURE 5.1
A two-body mass-spring system.

Let us now consider another choice for the state variables by selecting for the second body the position and velocity relative to those of the first body, rather than measured in absolute coordinates, that is

$$\begin{aligned}
\tilde{x}_1 &= x_1 \\
\tilde{x}_2 &= x_2 - x_1 \\
\tilde{x}_3 &= x_3 \\
\tilde{x}_4 &= x_4 - x_3
\end{aligned} \tag{5.10}$$

The new variables are related to the original ones by the relationship (5.11) with T^{-1} given by

$$T^{-1} = \begin{bmatrix} 1 & 0 & 0 & 0 \\ -1 & 1 & 0 & 0 \\ 0 & 0 & 1 & 0 \\ 0 & 0 & -1 & 1 \end{bmatrix} \tag{5.11}$$

The state matrices of the new representation are (they can be calculated by applying (5.7) or by rewriting the system equations using (5.10))

$$\tilde{A} = \begin{bmatrix} 0 & 0 & 1 & 0 \\ 0 & 0 & 0 & 1 \\ 0 & \frac{k}{m_1} & 0 & 0 \\ 0 & -\frac{k(m_1+m_2)}{m_1 m_2} & 0 & 0 \end{bmatrix} ; \quad \tilde{B} = \begin{bmatrix} 0 & 0 \\ 0 & 0 \\ \frac{1}{m_1} & 0 \\ -\frac{1}{m_1} & \frac{1}{m_2} \end{bmatrix} ; \tag{5.12}$$

$$\tilde{C} = \begin{bmatrix} 1 & 0 & 0 & 0 \end{bmatrix} ; \quad \tilde{D} - 0$$

5.2   Lagrange's formula

Assume that system (5.1) is at time $t_0 = 0$ in the state $\mathbf{x}(0) = \mathbf{x}_0$, then the time evolution of the state variables, namely $\mathbf{x}(t)$ $\forall t \geq 0$, can be calculated through the Lagrange's formula:

$$\mathbf{x}(t) = e^{At}\mathbf{x}_0 + \int_0^t e^{A(t-\tau)}\mathbf{Bu}(\tau)d\tau \tag{5.13}$$

and similarly the output $y(t)$ is given by

$$\mathbf{y}(t) = Ce^{At}\mathbf{x}_0 + \int_0^t Ce^{A(t-\tau)}\mathbf{Bu}(\tau)d\tau + \mathbf{Du}(t) \tag{5.14}$$

Here, the exponential matrix e^{At} is defined as:

$$e^{At} = \sum_{i=0}^{\infty} \frac{A^i t^i}{i!} \tag{5.15}$$

with $A^0 = I$.

The Lagrange's formula for the state can be rewritten as the sum of two terms:

$$\mathbf{x}(t) = \mathbf{x}_n(t) + \mathbf{x}_f(t) \tag{5.16}$$

The first term represents the natural evolution of the state

$$\mathbf{x}_n(t) = e^{At}\mathbf{x}_0 \tag{5.17}$$

while the second one the forced evolution

$$\mathbf{x}_f(t) = \int_0^t e^{A(t-\tau)}\mathbf{Bu}(\tau)d\tau \tag{5.18}$$

Analogously, the Lagrange's formula for the output consists of the two terms:

$$\mathbf{y}(t) = \mathbf{y}_n(t) + \mathbf{y}_f(t) \tag{5.19}$$

The first term is the natural response

$$\mathbf{y}_n(t) = Ce^{At}\mathbf{x}_0 \tag{5.20}$$

while the second is the forced response

$$\mathbf{y}_f(t) = C\int_0^t e^{A(t-\tau)}\mathbf{Bu}(\tau)d\tau + \mathbf{Du}(t) \tag{5.21}$$

5.3   Superposition principle

Suppose to calculate the movement $\mathbf{x}'(t)$ for system (5.1) with initial conditions $\mathbf{x}'_0$ and input $\mathbf{u}'(t)$. Suppose also to calculate the movement $\mathbf{x}''(t)$ for the same system, but this time starting from initial conditions $\mathbf{x}''_0$ and subject to the input $\mathbf{u}''(t)$. Consider then the same system with initial conditions $\alpha\mathbf{x}'_0 + \beta\mathbf{x}''_0$ and input $\alpha\mathbf{u}'(t) + \beta\mathbf{u}''(t)$ $(\alpha, \beta \in \mathbb{R})$. The time evolution $\mathbf{x}'''(t)$ of the system corresponding to this initial condition and input will be given by

$$
\begin{aligned}
\mathbf{x}'''(t) &= e^{At}(\alpha\mathbf{x}'_0 + \beta\mathbf{x}''_0) + \int_0^t e^{A(t-\tau)}B(\alpha\mathbf{u}'(\tau) + \beta\mathbf{u}''(\tau))d\tau \\
&= \alpha\left(e^{At}\mathbf{x}'_0 + \int_0^t e^{A(t-\tau)}B\mathbf{u}'(\tau)d\tau\right) + \beta\left(e^{At}\mathbf{x}''_0 + \int_0^t e^{A(t-\tau)}B\mathbf{u}''(\tau)d\tau\right) \\
&= \alpha\mathbf{x}'(t) + \beta\mathbf{x}''(t)
\end{aligned}
$$

$$(5.22)$$

Similarly, for the output we have that

$$\mathbf{y}'''(t) = \alpha\mathbf{y}'(t) + \beta\mathbf{y}''(t) \qquad (5.23)$$

The relationships $\mathbf{x}'''(t) = \alpha\mathbf{x}'(t) + \beta\mathbf{x}''(t)$ and $\mathbf{y}'''(t) = \alpha\mathbf{y}'(t) + \beta\mathbf{y}''(t)$ represent the *superposition principle*, namely, the property that, for any pair of values of α and β, the time evolution the state and of the output of the system starting from an initial condition given by $\alpha\mathbf{x}'_0 + \beta\mathbf{x}''_0$ and subject to the input $\alpha\mathbf{u}'(t) + \beta\mathbf{u}''(t)$ can be obtained by evaluating separately the effect of the initial condition $\mathbf{x}'_0$ and the input $\mathbf{u}'(t)$ and the effect of the initial condition $\mathbf{x}''_0$ and the input $\mathbf{u}''(t)$, and then considering the sum of the two effects, weighted by α and β.

5.4   The natural evolution

Let us consider the first of Eqs. (5.1) with $\mathbf{u} = 0$, namely $\dot{\mathbf{x}} = A\mathbf{x}$, and let us apply the Laplace transform to the left and right sides. We obtain

$$s\mathbf{X}(s) - \mathbf{x}_0 = A\mathbf{X}(s) \qquad (5.24)$$

which yields

$$\mathbf{X}(s) = (sI - A)^{-1}\mathbf{x}_0 \qquad (5.25)$$

On the other hand, considering $\mathbf{u} = 0$ in the Lagrange's formula (5.13), we have that

$$\mathbf{x}(t) = e^{At}\mathbf{x}_0 \qquad (5.26)$$

Comparing the two expressions (5.25) and (5.26), we immediately derive that

$$\mathcal{L}\left(e^{At}\right) = (sI - A)^{-1} \tag{5.27}$$

This equation provides a method to calculate the exponential matrix e^{At} as the inverse Laplace transform of the matrix $(sI - A)^{-1}$ (the inverse Laplace transform of a matrix is obtained by applying the inverse Laplace transform element-by-element to all the matrix coefficients). In the following examples, we use this result to calculate the natural evolution of a few systems.

Example 5.2 _____

Given $\mathbf{x}_0 = \begin{bmatrix} 0.1 \\ 0.4 \end{bmatrix}$, calculate the natural evolution of an LTI system with the following state matrices:

$$A = \begin{bmatrix} -1 & 0 \\ 0 & 3 \end{bmatrix}; \quad B = \begin{bmatrix} 1 \\ 1 \end{bmatrix};$$
$$C = \begin{bmatrix} 1 & 1 \end{bmatrix}; \quad D = 0 \tag{5.28}$$

We have that

$$sI - A = \begin{bmatrix} s+1 & 0 \\ 0 & s-3 \end{bmatrix} \tag{5.29}$$

Hence

$$(sI - A)^{-1} = \begin{bmatrix} \frac{1}{s+1} & 0 \\ 0 & \frac{1}{s-3} \end{bmatrix} \tag{5.30}$$

By calculating the inverse Laplace transform of $(sI - A)^{-1}$, we obtain the exponential matrix e^{At}:

$$e^{At} = \begin{bmatrix} e^{-t} & 0 \\ 0 & e^{3t} \end{bmatrix} \tag{5.31}$$

Finally, the natural evolution is given by

$$\mathbf{x}(t) = e^{At}\mathbf{x}_0 = \begin{bmatrix} e^{-t} & 0 \\ 0 & e^{3t} \end{bmatrix} \begin{bmatrix} 0.1 \\ 0.4 \end{bmatrix} = \begin{bmatrix} 0.1e^{-t} \\ 0.4e^{3t} \end{bmatrix} \tag{5.32}$$

Example 5.3 _____

Given $\mathbf{x}_0 = \begin{bmatrix} 1 \\ 2 \end{bmatrix}$, calculate the natural evolution of an LTI system with the following state matrices:

$$A = \begin{bmatrix} 0 & 1 \\ -2 & -3 \end{bmatrix}; \quad B = \begin{bmatrix} 1 \\ 1 \end{bmatrix};$$
$$C = \begin{bmatrix} 1 & 1 \end{bmatrix}; \quad D = 0 \tag{5.33}$$

We have that

$$sI - A = \begin{bmatrix} s & -1 \\ 2 & s+3 \end{bmatrix} \tag{5.34}$$

Since $\det(sI - A) = s^2 + 3s + 2$, the eigenvalues of the matrix A are $\lambda_1 = -1$, and $\lambda_2 = -2$.

Let us now calculate $(s\mathrm{I} - \mathrm{A})^{-1}$:

$$(s\mathrm{I} - \mathrm{A})^{-1} = \begin{bmatrix} \frac{s+3}{(s+1)(s+2)} & \frac{1}{(s+1)(s+2)} \\ \frac{-2}{(s+1)(s+2)} & \frac{s}{(s+1)(s+2)} \end{bmatrix} \tag{5.35}$$

We can expand the coefficients of this matrix in simple fractions as follows:

$$\begin{aligned} \frac{s+3}{(s+1)(s+2)} &= \frac{2}{s+1} - \frac{1}{s+2} \\ \frac{1}{(s+1)(s+2)} &= \frac{1}{s+1} - \frac{1}{s+2} \\ \frac{-2}{(s+1)(s+2)} &= -\frac{2}{s+1} + \frac{2}{s+2} \\ \frac{s}{(s+1)(s+2)} &= -\frac{1}{s+1} + \frac{2}{s+2} \end{aligned} \tag{5.36}$$

By calculating the inverse Laplace transform of $(s\mathrm{I} - \mathrm{A})^{-1}$, we obtain the exponential matrix $e^{\mathrm{A}t}$:

$$e^{\mathrm{A}t} = \begin{bmatrix} 2e^{-t} - e^{-2t} & e^{-t} - e^{-2t} \\ -2e^{-t} + 2e^{-2t} & -e^{-t} + 2e^{-2t} \end{bmatrix} \tag{5.37}$$

Finally, the natural evolution is given by

$$\mathbf{x}(t) = e^{\mathrm{A}t}\mathbf{x}_0 = \begin{bmatrix} 2e^{-t} - e^{-2t} & e^{-t} - e^{-2t} \\ -2e^{-t} + 2e^{-2t} & -e^{-t} + 2e^{-2t} \end{bmatrix} \begin{bmatrix} 1 \\ 2 \end{bmatrix} = \begin{bmatrix} 4e^{-t} - 3e^{-2t} \\ -4e^{-t} + 6e^{-2t} \end{bmatrix} \tag{5.38}$$

Example 5.4

Given $\mathbf{x}_0 = \begin{bmatrix} 2 \\ 1 \end{bmatrix}$, calculate the natural evolution of an LTI system with the following state matrices:

$$\begin{aligned} \mathrm{A} &= \begin{bmatrix} -3 & -1 \\ 1 & -3 \end{bmatrix}; & \mathrm{B} &= \begin{bmatrix} 1 \\ 1 \end{bmatrix}; \\ \mathrm{C} &= \begin{bmatrix} 1 & 1 \end{bmatrix}; & \mathrm{D} &= 0 \end{aligned} \tag{5.39}$$

We have that

$$s\mathrm{I} - \mathrm{A} = \begin{bmatrix} s+3 & 1 \\ -1 & s+3 \end{bmatrix} \tag{5.40}$$

Since $\det(s\mathrm{I} - \mathrm{A}) = s^2 + 6s + 10$, the eigenvalues of A are complex and conjugates: $\lambda_{1,2} = -3 \pm j$.

For this system, $(s\mathrm{I} - \mathrm{A})^{-1}$ reads

$$(s\mathrm{I} - \mathrm{A})^{-1} = \begin{bmatrix} \frac{s+3}{(s+3)^2+1} & \frac{-1}{(s+3)^2+1} \\ \frac{1}{(s+3)^2+1} & \frac{s+3}{(s+3)^2+1} \end{bmatrix} \tag{5.41}$$

By calculating the inverse Laplace transform of $(s\mathrm{I} - \mathrm{A})^{-1}$, we obtain the exponential matrix $e^{\mathrm{A}t}$:

$$e^{\mathrm{A}t} = \begin{bmatrix} e^{-3t}\cos t & -e^{-3t}\sin t \\ e^{-3t}\sin t & e^{-3t}\cos t \end{bmatrix} \tag{5.42}$$

Finally, the natural evolution is given by

$$\mathbf{x}(t) = e^{\mathrm{A}t}\mathbf{x}_0 = \begin{bmatrix} e^{-3t}\cos t & -e^{-3t}\sin t \\ e^{-3t}\sin t & e^{-3t}\cos t \end{bmatrix} \begin{bmatrix} 2 \\ 1 \end{bmatrix} = \begin{bmatrix} 2e^{-3t}\cos t - e^{-3t}\sin t \\ 2e^{-3t}\sin t + e^{-3t}\cos t \end{bmatrix} \tag{5.43}$$

Example 5.5

Given $\mathbf{x}_0 = \begin{bmatrix} 1 \\ 1 \end{bmatrix}$, calculate the natural evolution of an LTI system with the following state matrices:

$$A = \begin{bmatrix} -5 & 1 \\ 0 & -5 \end{bmatrix}; \quad B = \begin{bmatrix} 1 \\ 1 \end{bmatrix};$$
$$C = \begin{bmatrix} 1 & 1 \end{bmatrix}; \quad D = 0 \tag{5.44}$$

We have that

$$sI - A = \begin{bmatrix} s+5 & -1 \\ 0 & s+5 \end{bmatrix} \tag{5.45}$$

For this system, $(sI - A)^{-1}$ reads

$$(sI - A)^{-1} = \begin{bmatrix} \frac{1}{s+5} & \frac{1}{(s+5)^2} \\ 0 & \frac{1}{s+5} \end{bmatrix} \tag{5.46}$$

By calculating the inverse Laplace transform of $(sI - A)^{-1}$, we obtain the exponential matrix e^{At}:

$$e^{At} = \begin{bmatrix} e^{-5t} & te^{-5t} \\ 0 & e^{-5t} \end{bmatrix} \tag{5.47}$$

Finally, the natural evolution is given by

$$\mathbf{x}(t) = e^{At}\mathbf{x}_0 = \begin{bmatrix} e^{-5t} + te^{-5t} \\ e^{-5t} \end{bmatrix} \tag{5.48}$$

From these examples, we conclude that the natural evolution is the linear combination of elementary functions of the following types: $e^{\lambda_i t}$ for a simple and real eigenvalue λ_i; $t^{\eta_i-1}e^{\lambda_i t}$ with $\eta_i = 2, \ldots, k_i$ for a real eigenvalue λ_i of multiplicity k_i; $e^{\sigma_i t}\sin\omega_i t$, $e^{\sigma_i t}\cos\omega_i t$ for a pair of simple complex eigenvalues $\lambda_i = \sigma_i \pm \omega_i$; $t^{\eta_i-1}e^{\sigma_i t}\sin\omega_i t$, $t^{\eta_i-1}e^{\sigma_i t}\cos\omega_i t$ with $\eta_i = 2, \ldots, k_i$ for a pair of complex eigenvalues $\lambda_i = \sigma_i \pm \omega_i$ of multiplicity k_i. These elementary functions are called the *system modes* and determine the properties of the natural evolution of the system, as we will discuss in the next lectures.

We conclude the section with a practical example where we analyze the pendulum described in Sec. 2.4 for given values of its parameters l and m.

Example 5.6

Consider a pendulum, as in Fig. 2.4 (see Sec. 2.4), with length $l = 1$ m and mass $m = 0.5$ kg, and let $\omega = \sqrt{g/l} = \sqrt{9.8} \simeq 3$ rad/s. Suppose that the torque T_c is zero, so that the pendulum is in natural evolution. Assume also that the initial velocity is zero, and the initial angle is $5°$, that is $\mathbf{x}_0 = \begin{bmatrix} 5 \\ 0 \end{bmatrix}$.

Taking into account that for the pendulum the matrix A is given by $A = \begin{bmatrix} 0 & 1 \\ -\frac{g}{l} & 0 \end{bmatrix} = \begin{bmatrix} 0 & 1 \\ -\omega^2 & 0 \end{bmatrix}$, we have that

$$(sI - A)^{-1} = \begin{bmatrix} \frac{s}{s^2+\omega^2} & -\frac{1}{s^2+\omega^2} \\ \frac{\omega^2}{s^2+\omega^2} & \frac{s}{s^2+\omega^2} \end{bmatrix} \tag{5.49}$$

By calculating the inverse Laplace transform of $(sI - A)^{-1}$, we obtain the exponential matrix e^{At}:

$$e^{At} = \begin{bmatrix} \cos(\omega t) & \frac{1}{\omega}\sin(\omega t) \\ -\omega\sin(\omega t) & \cos(\omega t) \end{bmatrix} \tag{5.50}$$

Finally, the natural evolution is given by

$$\mathbf{x}(t) = e^{At}\mathbf{x}_0 = \begin{bmatrix} 5\cos(\omega t) \\ -5\omega\sin(\omega t) \end{bmatrix} \tag{5.51}$$

In particular, this shows that $\theta(t) = 5\cos(\omega t)$, namely the pendulum will oscillate between the two minimum and maximum angles of $\pm 5°$ with frequency $\omega \simeq 3$ rad/s and period $T = \frac{2\pi}{\omega} \simeq 2$ s. Notice that the frequency and the period do not depend on the pendulum mass.

5.5   The forced evolution

The forced evolution and the forced response of an LTI system are given by Eqs. (5.18) and (5.21). Let us now consider the case of SISO systems and assume as input the unit impulse $\delta(t)$, then Eqs. (5.18) and (5.21) become

$$h_x(t) = e^{At}B \tag{5.52}$$

and

$$h(t) = Ce^{At}B + D\delta(t) \tag{5.53}$$

Notice that these motions are equivalent to the natural evolution and response obtained starting from $\mathbf{x}(0) = B$.

The impulse response is particularly important as any forced evolution and response can be calculated from it. In fact

$$\mathbf{x}_f(t) = \int_0^t e^{A(t-\tau)}Bu(\tau)d\tau = \int_0^t h_x(t-\tau)u(\tau)d\tau = h_x(t) * u(t) \tag{5.54}$$

and

$$y_f(t) = \int_0^t \left(Ce^{A(t-\tau)}B + D\delta(t-\tau) \right) u(\tau)d\tau = \int_0^t h(t-\tau)u(\tau)d\tau = h(t) * u(t) \tag{5.55}$$

Eq. (5.54) shows that the forced evolution can be obtained as the convolution of the impulse evolution with the input signal. Similarly, Eq. (5.55) shows that the forced response can be obtained as the convolution of the impulse response with the input signal.

5.6   Symbolic calculation of the natural evolution of a system in MATLAB

Consider the LTI system of Example 5.3, with state matrices as in Eqs. (5.33). In this section, using MATLAB®, we calculate the system natural evolution from the initial conditions specified in Example 5.3.

First, we define a symbolic variable s with the following commands:

```
syms s
```

Next, we define the matrix A:

```
A=[0 1; -2 -3]
```

and, then, the matrix $(s\mathrm{I} - \mathrm{A})^{-1}$

```
AA=inv(s*eye(length(A))-A);
```

We are now ready to compute the Laplace antitransform of $(s\mathrm{I} - \mathrm{A})^{-1}$ using the command `ilaplace`:

```
exponentialmatrix=ilaplace(AA)
```

Lastly, we calculate the natural evolution starting from the given initial conditions

```
exponentialmatrix*[1 2]'
```

The result matches the natural evolution calculated in Example 5.3 (see Eq. (5.38)).

5.7   Numerical calculation of the natural and forced response in MATLAB

In this section, we numerically calculate the natural and forced response of a system, using MATLAB®. We exemplify the steps focusing on the linear pendulum in Eqs. (2.19) with state matrices in Eqs. (2.20). The parameters are fixed as: $m = 2$ kg and $l = 5$ m. We consider that the system is forced by the input $u(t) = \mathrm{step}(t)$.

First, we define the state matrices:

```
l=5;
m=2;
g=9.8;
A=[0 1; -g/l 0];
B=[0; 1/(m*l^2)];
C=[1 0];
D=0;
pendulumsys=ss(A,B,C,D);
```

We also define the initial conditions

```
x0=[0.1; 0]
```

corresponding to $\theta(0) = 0.1$ and $\omega(0) = 0$.

Next, we calculate and plot the natural evolution using the command initial:

```
T1=[0:0.01:200];
[ynat T]=initial(pendulumsys,x0,T1);
figure,plot(T1,ynat)
```

Notice that we have also specified the vector of the times where the free evolution is computed (the vector T1).

Next, we calculate and plot the forced response. To this aim, we use the command step:

```
[yforced T]=step(pendulumsys,T1);
figure,plot(T1,yforced)
```

Lastly, we calculate the complete response using the command lsim, which requires that one specifies the system to simulate, the input (given as an array of points), the time vector, and the initial condition:

```
U=ones(length(T1),1);
ycomplete=lsim(pendulumsys,U,T1,x0);
```

The complete response can be compared with the one obtained by summing the natural and forced term, as follows:

```
figure,plot(T1,ycomplete)
hold on
plot(T1,ynat+yforced,'r')
```

The comparison shows that the two curves are identical, due to the fact that the system is linear and, so, the superposition principle holds.

5.8   Exercises

1. Calculate the free evolution from initial conditions $\mathbf{x}_0 = \begin{bmatrix} 1 \\ 1 \end{bmatrix}$ for the LTI system with the following state matrices:

$$A = \begin{bmatrix} -3 & 1 \\ -1 & -3 \end{bmatrix}; \quad B = \begin{bmatrix} 1 \\ 1 \end{bmatrix}; \quad (5.56)$$
$$C = \begin{bmatrix} 1 & 0 \end{bmatrix}; \quad D = 0$$

2. Calculate the free evolution from initial conditions $\mathbf{x}_0 = \begin{bmatrix} 0.2 \\ 1 \end{bmatrix}$ for the LTI system with the following state matrices:

$$A = \begin{bmatrix} -3 & 1 \\ 0 & -3 \end{bmatrix}; \quad B = \begin{bmatrix} 1 \\ 1 \end{bmatrix};$$
$$C = \begin{bmatrix} 1 & 0 \end{bmatrix}; \quad D = 0 \tag{5.57}$$

3. Calculate the free evolution from initial conditions $\mathbf{x}_0 = \begin{bmatrix} -1 \\ 0.4 \end{bmatrix}$ for the LTI system with the following state matrices:

$$A = \begin{bmatrix} 0 & 1 \\ -1 & -1 \end{bmatrix}; \quad B = \begin{bmatrix} 0 \\ 1 \end{bmatrix};$$
$$C = \begin{bmatrix} 1 & 2 \end{bmatrix}; \quad D = 0 \tag{5.58}$$

4. Calculate the free evolution from initial conditions $\mathbf{x}_0 = \begin{bmatrix} 1 \\ 3 \\ -1 \end{bmatrix}$ for the LTI system with the following state matrices:

$$A = \begin{bmatrix} -3 & 0 & 0 \\ 0 & 2 & 0 \\ 0 & 0 & 1 \end{bmatrix}; \quad B = \begin{bmatrix} 1 \\ 1 \\ 1 \end{bmatrix};$$
$$C = \begin{bmatrix} 1 & 0 & 0 \end{bmatrix}; \quad D = 0 \tag{5.59}$$

5. Calculate the free evolution from initial conditions $\mathbf{x}_0 = \begin{bmatrix} 1 \\ -2 \end{bmatrix}$ for the LTI system with the following state matrices:

$$A = \begin{bmatrix} 0 & 1 \\ -6 & -7 \end{bmatrix}; \quad B = \begin{bmatrix} 2 \\ 2 \end{bmatrix};$$
$$C = \begin{bmatrix} 3.5 & 0.5 \end{bmatrix}; \quad D = 0 \tag{5.60}$$

6

Lecture 6—Equilibrium points and stability of continuous linear time-invariant systems

---

In this lecture, we introduce the notion of equilibrium points and that of stability for LTI systems. Then, we discuss the Routh criterion, namely a simple criterion for the stability of a continuous-time LTI system from its characteristic polynomial.

6.1 Equilibrium points

We continue our study of continuous-time LTI systems (5.1) by considering a special solution that is constant in time, namely $\mathbf{x}(t) = \bar{\mathbf{x}}$. This solution occurs for a constant input $\mathbf{u}(t) = \bar{\mathbf{u}}$. Since the solution is constant, its time derivative is zero. Hence $\bar{\mathbf{x}}$ is such that

$$0 = A\bar{\mathbf{x}} + B\bar{\mathbf{u}} \tag{6.1}$$

If A is invertible, namely if $\det A \neq 0$, then, in correspondence of the input $\bar{\mathbf{u}}$, there exists a unique equilibrium point $\bar{\mathbf{x}}$ given by

$$\bar{\mathbf{x}} = -A^{-1}B\bar{\mathbf{u}} \tag{6.2}$$

Otherwise, i.e., when $\det A = 0$, there are two possible scenarios: no solution exists or Eq. (6.1) has infinite solutions.

Example 6.1 ————————————————————————————

Consider an LTI system with the following state matrices:

$$A = \begin{bmatrix} 0 & 0 \\ 1 & -5 \end{bmatrix}; \quad B = \begin{bmatrix} 1 \\ 1 \end{bmatrix};$$
$$C = \begin{bmatrix} 1 & 1 \end{bmatrix}; \quad D = 0 \tag{6.3}$$

Then, if $\bar{u} \neq 0$, no solution exists. If $\bar{u} = 0$, then all points with $\bar{x}_1 = 5\bar{x}_2$ are equilibrium points for the linear system.

DOI: 10.1201/9781003487289-6

6.2 Stability of continuous linear time-invariant systems

Equilibrium points may be stable, asymptotic stable or unstable. This depends on the behavior that is obtained when the motion is perturbed. In more detail, the equilibrium point can be viewed as the nominal motion, namely the motion obtained by the nominal conditions, i.e., initial conditions $\mathbf{x}_0 = \bar{\mathbf{x}}$ and input $\bar{\mathbf{u}}$, while any motion that is obtained with the same input $\bar{\mathbf{u}}$, but different initial conditions $\mathbf{x}_0 \neq \bar{\mathbf{x}}$, can be considered as a perturbed motion.

The equilibrium point is said to be stable, if for any $\varepsilon > 0$ there exists $\delta > 0$, such that selecting the initial conditions close enough to the equilibrium point, namely such that $\|\mathbf{x}_0 - \bar{\mathbf{x}}\| < \delta$, then the perturbed motion $\mathbf{x}(t)$, which starts from such initial conditions, remains close enough to the equilibrium point: $\|\mathbf{x}(t) - \bar{\mathbf{x}}\| < \varepsilon$.

The stability is asymptotic, if in addition to the previous conditions, it holds that

$$\lim_{t \to +\infty} \mathbf{x}(t) = \bar{\mathbf{x}} \tag{6.4}$$

If the condition for stability does not hold, then the equilibrium point is said to be unstable.

Let us now consider the difference between the perturbed motion ($\mathbf{x}(t)$ starting from $\mathbf{x}_0$) and the nominal one (i.e., the equilibrium $\bar{\mathbf{x}}$): $\delta\mathbf{x} = \mathbf{x}(t) - \bar{\mathbf{x}}$. By differentiation, and taking into account that the input is the same for the two motions, we derive that

$$\dot{\delta\mathbf{x}} = A\delta\mathbf{x} \tag{6.5}$$

The point is stable if the solutions of (6.5) are bounded, it is asymptotic stable if these solutions tend to zero, and it is unstable if they are not bounded.

Also notice that the solutions of (6.5) are the natural evolution of the original linear system. Finally, we notice that these solutions do not depend on the equilibrium point, such that stability is a property of the whole system.

Taking into account all previous considerations, we conclude that an LTI system is asymptotically stable if and only if:

$$\lim_{t \to +\infty} e^{At} = 0 \tag{6.6}$$

The stability of an LTI system only depends on the state matrix A. More specifically, we now show that it only depends on the eigenvalues of A.

In fact, we have seen in the previous lecture that the modes that appear in the exponential matrix are of the type: $e^{\lambda_i t}$, $t^{\eta_i - 1} e^{\lambda_i t}$, $e^{\sigma_i t} \sin \omega_i t$, $e^{\sigma_i t} \cos \omega_i t$, $t^{\eta_i - 1} e^{\sigma_i t} \sin \omega_i t$, $t^{\eta_i - 1} e^{\sigma_i t} \cos \omega_i t$, ... Hence, if (and only if) all eigenvalues have negative real part, all modes go to zero for $t \to +\infty$, and so does e^{At}.

We conclude that a system is asymptotically stable if and only if all eigenvalues of A have negative real part.

If at least one eigenvalue of A has a positive real part, then the system is unstable.

The system is stable if all eigenvalues have non-positive real part and the eigenvalues on the imaginary axis are simple. If the eigenvalues on the imaginary axis have a multiplicity greater than one, then it is important to find which modes associated with these eigenvalues are present in e^{At} (this depends on the geometric multiplicity of these eigenvalues). We do not delve into this topic but limit the discussion to exemplify the two cases that can appear.

Example 6.2 _____

Consider an LTI system with the following state matrix $A = A_1 = \begin{bmatrix} 0 & 0 \\ 0 & 0 \end{bmatrix}$, and a second LTI system with $A = A_2 = \begin{bmatrix} 0 & 1 \\ 0 & 0 \end{bmatrix}$.

For the first system, we get

$$e^{A_1 t} = \begin{bmatrix} \text{step}(t) & 0 \\ 0 & \text{step}(t) \end{bmatrix} \tag{6.7}$$

while for the second system we obtain

$$e^{A_2 t} = \begin{bmatrix} \text{step}(t) & t \\ 0 & \text{step}(t) \end{bmatrix} \tag{6.8}$$

We conclude that the first system is stable (but not asymptotically), while the second system is unstable because of the presence of the mode t in the exponential matrix.

Let us now consider a few examples of stability analysis from the knowledge of the system's state matrix A.

Example 6.3 _____

Consider an LTI system with $A = \begin{bmatrix} -1 & 1 \\ 0 & 0.5 \end{bmatrix}$.

The matrix A is upper triangular. The eigenvalues are on the diagonal. They are $\lambda_1 = -1$ and $\lambda_2 = 0.5$. Since λ_2 is positive, the system is unstable.

Example 6.4 _____

Consider the pendulum described by Eqs. (2.20).

Since the system state matrix A is given by $A = \begin{bmatrix} 0 & 1 \\ -\frac{g}{l} & 0 \end{bmatrix}$, the system eigenvalues are: $\lambda_{1,2} = \pm j\sqrt{\frac{g}{l}}$. They are on the imaginary axis and are simple. Hence, the system is stable but not asymptotically stable.

Example 6.5 _____

Consider now the inverted pendulum described by Eqs. (2.24).

In this case, the matrix A is given by $A = \begin{bmatrix} 0 & 1 \\ \frac{g}{l} & 0 \end{bmatrix}$, and the system eigenvalues are: $\lambda_{1,2} = \pm\sqrt{\frac{g}{l}}$. One of them is positive and, hence, the system is unstable.

Example 6.6 _____

Consider an LTI system with $A = \begin{bmatrix} -5 & -1 & 1 \\ 0 & 0 & 1 \\ 0 & -2 & -2 \end{bmatrix}$.

The characteristic polynomial of A is: $p(\lambda) = (\lambda+5)(\lambda^2+2\lambda+2)$. The system eigenvalues are: $\lambda_1 = -5$, $\lambda_{2,3} = -1 \pm j$. All of them lie on the open left half-plane, so the system is asymptotically stable.

Example 6.7 _____

Consider an LTI system with $A = \begin{bmatrix} 0 & -1 & 1 \\ 0 & 0 & 1 \\ 0 & -2 & -2 \end{bmatrix}$.

The characteristic polynomial of A is: $p(\lambda) = \lambda(\lambda^2 + 2\lambda + 2)$. The system eigenvalues are: $\lambda_1 = 0$, $\lambda_{2,3} = -1 \pm j$. Two eigenvalues lie on the open left half-plane, one is on the origin, so the system is stable, but not asymptotically stable.

6.3   The Routh criterion

In the previous examples, we have determined whether a system is stable or not, by first calculating the characteristic polynomial of the state matrix A, and then calculating the roots of this polynomial, that is, the eigenvalues of A. However, this second step is not mandatory, as we do not need to know the exact values of the eigenvalues but only their position on the complex plane. To this aim, we can use the Routh criterion that from the analysis of a polynomial of order n makes possible to determine where the roots of the polynomial are located in the complex plane, more specifically how many of them are in the open left half-plane, how many on the imaginary axis, and how many in the open right half-plane.

Let us consider the following polynomial

$$p(\lambda) = a_n\lambda^n + a_{n-1}\lambda^{n-1} + a_{n-2}\lambda^{n-2} + \ldots + a_1\lambda + a_0 \qquad (6.9)$$

and indicate with n_n the number of roots with negative real part, n_o the number of roots that lie on the imaginary axis, and with n_p the number of roots with positive real part. Clearly $n_p + n_n + n_o = n$.

The first step of the Routh criterion is the construction of the so-called Routh table (Table 6.1). The table has $n + 1$ rows. The table is iteratively constructed, from the top to the bottom. The first two rows are constructed by placing the coefficients of the polynomial terms with the same oddity from the highest to the lowest power. Suppose that n is even, so the first row of the table (row labeled as n) will contain all coefficients of even power terms, while the next one (row labeled as $n - 1$) will contain all coefficients of odd power terms. Viceversa, if n is odd, the first row will contain the coefficients of odd power terms, and the second one the coefficients of even power terms.

Each row of the Routh table is computed from the two previous rows. Namely, to calculate row i, one has to use the coefficients appearing in rows $i + 1$ and $i + 2$. The generic term r_i appearing in this row is obtained as:

$$r_i = -\frac{\begin{vmatrix} p_{i+2} & p_i \\ q_{i+1} & q_{i-1} \end{vmatrix}}{q_{i+1}}, \text{ if } q_{i+1} \neq 0.$$

From this expression, it becomes clear that constructing all the rows of the Routh table is not always feasible. We can classify the situation into three distinct cases, based on whether or not the Routh table can be fully completed.

TABLE 6.1
Construction of the Routh table.

n	a_n	a_{n-2}	$\ldots$
n-1	a_{n-1}	a_{n-3}	$\ldots$
	$\vdots$		
i+2	p_{i+2}	p_i	$\ldots$
i+1	q_{i+1}	q_{i-1}	$\ldots$
i	r_i	$\ldots$	
	$\vdots$		

- Case 1. Completed Routh table.

  In this case, all coefficients of the Routh table have been calculated. This means that all terms in the first column are different than zeros. In this case, $n_0 = 0$. To compute n_n we count how many coefficients in successive rows have the same sign. Conversely, n_p corresponds to the number of times the coefficients in successive rows have opposite signs.

- Case 2. A coefficient of one row is zero, but all other coefficients are non-zero.

  In such cases the zero coefficient is replaced by a small positive quantity $\varepsilon > 0$. The other terms of the Routh table are calculated and then the analysis is done as for case 1, considering $\varepsilon \to 0^+$.

- Case 3. All coefficients of an odd row are equal to zero.

  In this case, the Routh table cannot be completed. We count how many changes of sign or permanences are found in the rows that have been completed. Finally, we derive the sign of the remaining roots by explicitly calculating them from the polynomial associated with the last non-zero row. In practice, if all elements of row i are zero, then we consider the polynomial $q(\lambda) = q_{i+1}\lambda^{i+1} + q_{i-1}\lambda^{i-1} + \ldots$ and find its roots $q(\lambda) = 0$. Given the form of $q(\lambda)$ these roots are symmetric with respect to both the $x-$ and the $y-$axis.

Let us now consider a few examples of the application of the Routh criterion to characteristic polynomials of the state matrix A.

Example 6.8

Consider the polynomial $p(\lambda) = \lambda^3 - 4\lambda^2 + \lambda + 6$. The Routh table for this polynomial is Table 6.2. This table is complete, so we are in case 1, and $n_0 = 0$. We now count the changes of signs in the first column between successive rows. We move from 1 to -4, and then from -4 to $\frac{5}{2}$, and finally from $\frac{5}{2}$ to 6. We count two changes of sign, so $n_p = 2$. Finally $n_n = n - n_p = 1$.

The roots of the polynomial can be calculated in MATLAB® with the command `roots`. The input parameter is a vector containing from the highest to the lowest power the coefficients of all terms of the polynomial (if a given power does not appear in the polynomial, we still need to explicitly write the coefficient, namely zero, as the order of the polynomial is equal to the number of elements of the vector, plus one). For this polynomial, hence, the roots can be obtained using `roots([1 -4 1 6])`. We find: $\lambda_1 = -1$, $\lambda_2 = 2$, and $\lambda_3 = 3$, which confirm the conclusion drawn with the Routh criterion.

TABLE 6.2
Routh table for Example 6.8.

3	1	1
2	-4	6
1	$\frac{5}{2}$	
0	6	

Example 6.9

Consider the polynomial $p(\lambda) = 4\lambda^4 + 3\lambda^3 + 5\lambda^2 + 2\lambda + 1$, whose Routh table is Table 6.3. The table is complete, so we are in case 1, and $n_0 = 0$. In the first column there are no changes of sign, so $n_p = 0$, and $n_n = n = 4$.

The roots of the polynomial can be calculated in MATLAB® with the command `roots([4 3 5 2 1])`. We get: $\lambda_{1,2} = -0.1152 \pm j0.9029$, and $\lambda_{3,4} = -0.2598 \pm j0.4840$.

TABLE 6.3
Routh table for Example 6.9.

4	4	5	1
3	3	2	
2	$\frac{7}{3}$	1	
1	$\frac{5}{7}$		
0	1		

Example 6.10

Consider now the polynomial $p(\lambda) = \lambda^3 + 3\lambda - 2$. The first two rows of the Routh table are illustrated in Table 6.4. The first coefficient of the second row is zero, but the other one is different than zero. We are thus in case 2. We now replace this zero coefficient and go ahead with the construction of the Routh table as in Table 6.5. We are able to complete the Routh table, and we study the signs of the coefficients of the first column, taking into account that $\frac{3\varepsilon+2}{\varepsilon} > 0$. We conclude that: $n_0 = 0$, $n_p = 1$, $n_n = 2$. Finally, we check the result by calculating in MATLAB® the roots of $p(\lambda)$ with the command `roots([1 0 3 -2])`. We get: $\lambda_1 = 0.5961$ and $\lambda_{2,3} = -0.2980 \pm j1.8073$.

TABLE 6.4

First two rows of the Routh table for Example 6.10.

3	1	3
2	0	-2

TABLE 6.5

Routh table for Example 6.10.

3	1	3
2	ε	-2
1	$\frac{3\varepsilon+2}{\varepsilon}$	
0	-2	

Example 6.11 _____

Consider the following polynomial: $p(\lambda) = \lambda^6 + \lambda^5 - 2\lambda^4 - 3\lambda^3 - 7\lambda^2 - 4\lambda - 4$. As shown in Table 6.6 rows labeled as 5 and 4 are equal, hence all coefficients of the row labeled as 3 are zeros. We are in case 3. Let us then consider the polynomial associated with the row labeled as 4: $q(\lambda) = \lambda^4 - 3\lambda^2 - 4$. Now let $z = \lambda^2$. Hence $q(z) = z^2 - 3z - 4$. $q(z) = 0$ yields $z_1 = 4$ and $z_2 = -1$. And so $\lambda_{1,2} = \pm 2$ and $\lambda_{3,4} = \pm j$. The remaining two roots of $p(\lambda)$ are positive since the coefficients of the first three rows all have the same sign. Using the MATLAB® command roots([1 1 -2 -3 -7 -4 -4]) we get beyond the four roots already calculated the following two roots: $\lambda_{5,6} = -0.5000 \pm j0.8660$.

TABLE 6.6

Routh table for Example 6.11.

6	1	-2	7	-4
5	1	-3	-4	
4	1	-3	-4	
3	0	0		

Example 6.12 _____

We now discuss an example with a characteristic polynomial that is a function of a parameter. In more detail, let us consider the polynomial $p(\lambda, k) = \lambda^3 + 5\lambda^2 + 2k\lambda + 1$, with $k \in \mathbb{R}$. The Routh table associated with this polynomial is given in Table 6.7, which shows that a coefficient of the first column is a function of k. For $k > 1/10$ all terms of the first column are positive and, hence, all roots of the polynomial have negative real part. For $k < 1/10$ we find two changes of sign in the first column, indicating that two roots of the polynomial have a positive real part. Lastly, for $k = 1/10$ a row of odd index ($i = 1$) becomes zero. Hence, we consider the polynomial $q(\lambda) = 5\lambda^2 + 1$ and calculate its roots. As $q(\lambda) = 0$ yields $\lambda_{1,2} = \pm\frac{1}{\sqrt{5}}$, we conclude that for $k = 1/10$ there are two purely imaginary roots and one with a negative real part (as the first two terms of the first column of the Routh Table 6.7 have the same sign).

TABLE 6.7
Routh table for Example 6.12.

3	1	2k
2	5	1
1	$\frac{10k-1}{5}$	
0	1	

MATLAB® exercise 6.1 _____

In the previous exercises we have seen that the MATLAB® command `roots` returns the roots of a polynomial. Here, we show a further example of the application of this command, in order to study a polynomial that is a function of a parameter. We refer to the polynomial analyzed in Example 6.12, namely $p(\lambda, k) = \lambda^3 + 5\lambda^2 + 2k\lambda + 1$. We start by defining a vector containing all values of k that we want to test:

```
kv=[-1:0.01:3]
```

Here, k is varied in the interval $[-1, 3]$ at steps of 0.01. Next, we write a `for` cycle where at each iteration, we calculate the roots of the polynomial for a given k and we plot them in the plane $(\text{Re}(\lambda), \text{Im}(\lambda))$.

```
figure
for j=1:length(kv)
    k=kv(j);
    calculatedroots=roots([1 5 2*k 1]);
    for i=1:3
        hold on
        plot(real(calculatedroots(i)),imag(calculatedroots(i)),'.')
    end
end
```

The result is illustrated in Fig. 6.1, which shows that two branches cross the imaginary axis, indicating that there are values of k for which two roots have positive real part. This is an example of a *root locus*, a powerful tool for the representation of the system eigenvalues when a parameter is varied. The root locus is not discussed in this book.

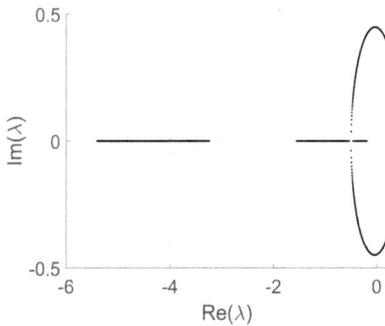

FIGURE 6.1
Positions of the roots of $p(\lambda, k) = \lambda^3 + 5\lambda^2 + 2k\lambda + 1$ when k is varied in $[-1, 3]$.

6.4 Exercises

1. For the system with state matrices in Eq. (5.57), calculate the equilibrium point corresponding to the input $\bar{u} = 3$.

2. Study the stability of the system with state matrices in Eq. (5.56).

3. Study the stability of the LTI system with the following state matrices:

$$A = \begin{bmatrix} -14 & -4 & -10 \\ -13 & -5 & -10 \\ 34 & 16 & 27 \end{bmatrix} ; \quad B = \begin{bmatrix} 1 \\ 1 \\ 1 \end{bmatrix} ; \quad (6.10)$$
$$C = \begin{bmatrix} 1 & 0 & 1 \end{bmatrix} ; \quad\quad\quad D = 2$$

4. Apply the Routh criterion to study the position in the complex plane of the roots of the polynomial $p(s) = s^6 + 4s^5 + 3s^4 + 16s^2 + 64s + 48$.

5. Apply the Routh criterion to study the position in the complex plane of the roots of the polynomial $p(s) = s^3 + 2s^2 + ks + 3$ with $k \in \mathbb{R}$.

7

Lecture 7—The transfer function

In this lecture, we introduce the notion of transfer function and its relationship with the state-space representation of a system. We refer to continuous time systems, although as we will see in the other lectures the notion is more general. Next, we discuss the concept of BIBO stability and how to study it.

7.1 Definition and invariance to state transformation

The transfer function (respectively, the transfer matrix) expresses the relationship between the Laplace transform of the input and that of the output of a SISO (respectively MIMO) system, for zero initial conditions.

Consider a continuous-time LTI system (5.1), then applying the Laplace transform to the left and right side of each equation we get

$$s\mathbf{X}(s) - \mathbf{x}(0) = \mathrm{A}\mathbf{X}(s) + \mathrm{B}\mathbf{U}(s)$$
$$\mathbf{Y}(s) = \mathrm{C}\mathbf{X}(s) + \mathrm{D}\mathbf{U}(s) \tag{7.1}$$

Taking into account $\mathbf{x}(0) = 0$, we obtain

$$(s\mathrm{I} - \mathrm{A})\,\mathbf{X}(s) = \mathrm{B}\mathbf{U}(s)$$
$$\mathbf{Y}(s) = \mathrm{C}\mathbf{X}(s) + \mathrm{D}\mathbf{U}(s) \tag{7.2}$$

From which we finally get:

$$\mathbf{Y}(s) = \left(\mathrm{C}\,(s\mathrm{I} - \mathrm{A})^{-1}\,\mathrm{B} + \mathrm{D}\right)\mathbf{U}(s) \tag{7.3}$$

The transfer function/matrix is given by

$$\mathrm{G}(s) = \mathrm{C}\,(s\mathrm{I} - \mathrm{A})^{-1}\,\mathrm{B} + \mathrm{D} \tag{7.4}$$

For a SISO system $G(s)$ is a scalar and can be defined as: $G(s) = \frac{Y(s)}{U(s)}$. For a MIMO system $\mathrm{G}(s) \in \mathbb{C}^{p \times m}$ and can be expressed as:

$$\mathrm{G}(s) = \begin{bmatrix} g_{11}(s) & \cdots & g_{1m}(s) \\ \vdots & & \\ g_{p1}(s) & \cdots & g_{pm}(s) \end{bmatrix} \tag{7.5}$$

DOI: 10.1201/9781003487289-7

The generic term of this matrix, namely $g_{ij}(s)$, represents the transfer function from input j to ouput i, assumed that all other inputs are set to zero: $g_{ij}(s) = \frac{Y_i(s)}{U_j(s)}$, with $u_h = 0$ for $h \neq j$.

The transfer matrix is invariant to state transformations of type (5.2). In fact, in the new reference system with variables $\tilde{\mathbf{x}}$, the transfer matrix reads

$$\tilde{G}(s) = \tilde{C} \left(s\mathbf{I} - \tilde{A}\right)^{-1} \tilde{B} + \tilde{D} \tag{7.6}$$

Let us now apply the relationships (5.7). We obtain

$$
\begin{aligned}
\tilde{G}(s) &= \mathbf{CT} \left(s\mathbf{I} - \mathbf{T}^{-1}\mathbf{AT}\right)^{-1} \mathbf{T}^{-1}\mathbf{B} + \mathbf{D} \\
&= \mathbf{CT} \left(s\mathbf{T}^{-1}\mathbf{T} - \mathbf{T}^{-1}\mathbf{AT}\right)^{-1} \mathbf{T}^{-1}\mathbf{B} + \mathbf{D} \\
&= \mathbf{CTT}^{-1} \left(s\mathbf{I} - \mathbf{A}\right)^{-1} \mathbf{TT}^{-1}\mathbf{B} + \mathbf{D} \\
&= \mathbf{C} \left(s\mathbf{I} - \mathbf{A}\right)^{-1} \mathbf{B} + \mathbf{D} \\
&= G(s)
\end{aligned}
\tag{7.7}
$$

Hence the transfer matrix does not depend on the reference system of the state-space representation.

The transfer function is so important that many times LTI systems are described using their transfer function.

The transfer function allows one to immediately calculate the output in response to a given input (in such cases, even if often omitted, one refers to the forced response, as the initial conditions are assumed to be zero when transfer functions are considered):

$$Y(s) = G(s)U(s) \tag{7.8}$$

A particular case is when $u(t) = \delta(t)$. In this case, $U(s) = 1$, and $Y(s) = G(s)$. This means that the transfer function is the Laplace transform of the impulsive response. This can be viewed also by considering Eq. (5.53). By applying the Laplace transform to $h(t) = \mathbf{C}e^{\mathbf{A}(t)}\mathbf{B} + \mathbf{D}\delta(t)$, we get $G(s) = \mathbf{C}(s\mathbf{I} - \mathbf{A})^{-1}\mathbf{B} + \mathbf{D}$, that is, Eq. (7.4).

7.2   Poles and zeros

For a SISO system the transfer function is the ratio of two polynomials in s:

$$G(s) = \frac{N(s)}{D(s)} \tag{7.9}$$

In this expression, there can be cancellations between one or more roots of the numerators and the corresponding ones in the denominators. When this happens, there are hidden variables that do not influence the input-output relationship of the system. On the contrary, when there are no cancellations, then the system is said to be in minimal form. In this case, the polynomial $D(s)$ is of order n, that is, the system order, while the one in the numerator, $N(s)$, is of order less or equal to n.

The roots of these two polynomials are important quantities. The roots of $D(s)$ are called the poles of the system, and the ones of $N(s)$ are called the zeros of the system. Thus a system has exactly n poles and up to n finite zeros, say $n_z \leq n$. The remaining $n - n_z$ are called infinite zeros. Systems with all poles and zeros with negative real part are called minimum phase systems. Consequently, a system that has one or more zeros or poles with positive real part is non-minimum phase.

For a square MIMO system $(m = p)$ the transfer matrix can be rewritten as follows:

$$G(s) = \frac{1}{D(s)} N(s) \tag{7.10}$$

where $D(s) = \det (sI - A)^{-1}$ and $N(s) \in \mathbb{C}^{m \times m}$. As for the SISO case, the poles are the roots of the polynomial $D(s)$. Instead, the zeros are the solutions of $\det N(s) = 0$.

Since the transfer function/matrix is invariant to state transformation, then the system poles and zeros are invariant quantities as well.

7.3   Examples of calculation of the transfer function

We illustrate now the calculation of the transfer function through a series of examples.

Example 7.1

Transfer function of the pendulum.
Let us consider the pendulum of Eq. (2.18). We recall here the state-space matrices that we have derived in Lecture 2. They are:

$$A = \begin{bmatrix} 0 & 1 \\ -\frac{g}{l} & 0 \end{bmatrix}; \quad B = \begin{bmatrix} 0 \\ \frac{1}{ml^2} \end{bmatrix}; \tag{7.11}$$
$$C = \begin{bmatrix} 1 & 0 \end{bmatrix}; \quad D = 0$$

Let us first compute $(sI - A)^{-1}$.
Since

$$sI - A = \begin{bmatrix} s & -1 \\ \frac{g}{l} & s \end{bmatrix} \tag{7.12}$$

and so

$$(sI - A)^{-1} = \begin{bmatrix} \frac{s}{s^2 + \frac{g}{l}} & \frac{1}{s^2 + \frac{g}{l}} \\ -\frac{\frac{g}{l}}{s^2 + \frac{g}{l}} & \frac{s}{s^2 + \frac{g}{l}} \end{bmatrix} \tag{7.13}$$

Finally, we calculate the transfer function $G(s)$ as

$$G(s) = C\,(sI - A)^{-1}\,B = \begin{bmatrix} 1 & 0 \end{bmatrix} \begin{bmatrix} \frac{s}{s^2+\frac{g}{l}} & \frac{1}{s^2+\frac{g}{l}} \\ -\frac{\frac{g}{l}}{s^2+\frac{g}{l}} & \frac{s}{s^2+\frac{g}{l}} \end{bmatrix} \begin{bmatrix} 0 \\ \frac{1}{ml^2} \end{bmatrix} = $$

$$= \begin{bmatrix} \frac{s}{s^2+\frac{g}{l}} & \frac{1}{s^2+\frac{g}{l}} \end{bmatrix} \begin{bmatrix} 0 \\ \frac{1}{ml^2} \end{bmatrix} = \frac{\frac{1}{ml^2}}{s^2+\frac{g}{l}} \qquad (7.14)$$

The same result can be obtained by starting from Eq. (2.18) that we now rewrite using $y = \theta$ and $u = T_c$:

$$\ddot{y} + \frac{g}{l}y = \frac{1}{ml^2}u \qquad (7.15)$$

By applying the Laplace transform to the left and right side of the equation, and considering that $y(0) = \dot{y}(0) = 0$, we get

$$s^2 Y(s) + \frac{g}{l}Y(s) = \frac{1}{ml^2}U(s) \qquad (7.16)$$

and so

$$G(s) = \frac{Y(s)}{U(s)} = \frac{\frac{1}{ml^2}}{s^2 + \frac{g}{l}} \qquad (7.17)$$

Example 7.2

Tranfer function of a second-order LTI system.
Let us consider again the LTI system with state matrices (5.39). We recall here these matrices for easiness of reference:

$$A = \begin{bmatrix} -3 & -1 \\ 1 & -3 \end{bmatrix}; \quad B = \begin{bmatrix} 1 \\ 1 \end{bmatrix};$$
$$C = \begin{bmatrix} 1 & 1 \end{bmatrix}; \qquad D = 0 \qquad (7.18)$$

In Lecture 5 we have found that for this system the matrix $(sI - A)^{-1}$ is given by

$$(sI - A)^{-1} = \begin{bmatrix} \frac{s+3}{(s+3)^2+1} & \frac{-1}{(s+3)^2+1} \\ \frac{1}{(s+3)^2+1} & \frac{s+3}{(s+3)^2+1} \end{bmatrix} \qquad (7.19)$$

Let us now compute the transfer function

$$G(s) = C\,(sI - A)^{-1}\,B = \begin{bmatrix} 1 & 1 \end{bmatrix} \begin{bmatrix} \frac{s+3}{(s+3)^2+1} & \frac{-1}{(s+3)^2+1} \\ \frac{1}{(s+3)^2+1} & \frac{s+3}{(s+3)^2+1} \end{bmatrix} \begin{bmatrix} 1 \\ 1 \end{bmatrix}$$

$$= \begin{bmatrix} \frac{s+4}{(s+3)^2+1} & \frac{s+2}{(s+3)^2+1} \end{bmatrix} \begin{bmatrix} 1 \\ 1 \end{bmatrix} = \frac{2(s+3)}{(s+3)^2+1} = \frac{2(s+3)}{s^2+6s+10} \qquad (7.20)$$

Example 7.3

Transfer function of the quarter-car model.
In order to derive the transfer function of the quarter-car model of Fig. 2.2, it is convenient to start from Eqs. (2.9) that we rewrite here considering $y = x_2$ and $u = r$:

$$b(\dot{y} - \dot{x}_1) + k_s(y - x_1) - k_w(x_1 - u) = m_1\ddot{x}_1$$
$$-k_s(y - x_1) - b(\dot{y} - \dot{x}_1) = m_2\ddot{y} \qquad (7.21)$$

We apply the Laplace transform to both sides of the equations, obtaining

$$bsY(s) + k_sY(s) + k_wU(s) = (m_1 s^2 + bs + k_s + k_w)\,X_1(s)$$
$$(m_2 s^2 + bs + k_s)\,Y(s) = (k_s + sb)X_1(s) \qquad (7.22)$$

From the first of these equations we get

$$X_1(s) = \frac{(bs + k_s)Y(s) + k_w U(s)}{(m_1 s^2 + bs + k_s + k_w)} \tag{7.23}$$

and replacing in the second equation

$$(m_2 s^2 + bs + k_s)(m_1 s^2 + bs + k_s + k_w)Y(s) = (k_s + sb)^2 Y(s) + (k_s + sb)k_w U(s) \tag{7.24}$$

and finally

$$G(s) = \frac{Y(s)}{U(s)} = \frac{(k_s + sb)k_w}{m_1 m_2 s^4 + b(m_1 + m_2)s^3 + (m_1 k_s + m_2(k_s + k_w))s^2 + b k_w s + k_w k_s} \tag{7.25}$$

Example 7.4

Tranfer function of an RLC circuit.
Consider the RLC circuit studied in Example 2.7, having the following state matrices:

$$A = \begin{bmatrix} -\frac{R}{L} & -\frac{1}{L} \\ \frac{1}{C} & 0 \end{bmatrix}; \quad B = \begin{bmatrix} \frac{1}{L} \\ 0 \end{bmatrix};$$
$$C = \begin{bmatrix} 0 & 1 \end{bmatrix}; \quad D = 0 \tag{7.26}$$

First, we calculate the matrix $(sI - A)^{-1}$:

$$(sI - A)^{-1} = \frac{1}{s^2 + \frac{R}{L}s + \frac{1}{LC}} \begin{bmatrix} s & -\frac{1}{L} \\ \frac{1}{C} & s + \frac{R}{L} \end{bmatrix} \tag{7.27}$$

Next, we compute the transfer function as follows:

$$\begin{aligned} G(s) &= C(sI - A)^{-1}B \\ &= \frac{1}{s^2 + \frac{R}{L}s + \frac{1}{LC}} \begin{bmatrix} \frac{1}{C} & s + \frac{R}{L} \end{bmatrix} \begin{bmatrix} s & -\frac{1}{L} \\ \frac{1}{C} & s + \frac{R}{L} \end{bmatrix} \begin{bmatrix} \frac{1}{L} \\ 0 \end{bmatrix} \\ &= \frac{\frac{1}{LC}}{s^2 + \frac{R}{L}s + \frac{1}{LC}} \end{aligned} \tag{7.28}$$

Example 7.5

Transfer function of an LTI system in diagonal form.
Consider the following LTI system in diagonal form, namely a SISO system with diagonal A:

$$A = \begin{bmatrix} \lambda_1 & & & \\ & \lambda_2 & & \\ & & \ddots & \\ & & & \lambda_n \end{bmatrix}; \quad B = \begin{bmatrix} b_1 \\ b_2 \\ \vdots \\ b_n \end{bmatrix};$$
$$C = \begin{bmatrix} c_1 & c_2 & \cdots & c_n \end{bmatrix}; \quad D = 0 \tag{7.29}$$

By direct calculation, we find that the transfer function is given by

$$G(s) = \frac{b_1 c_1}{s - \lambda_1} + \frac{b_2 c_2}{s - \lambda_2} + \ldots + \frac{b_n c_n}{s - \lambda_n} \tag{7.30}$$

This result is also important for obtaining a straightforward realization of a transfer function, specifically to find a set of state matrices A, B, C, and D that corresponds to the given transfer function. This approach is applicable when the transfer function has real and distinct poles. The transfer function is first rewritten as

$$G(s) = \frac{R_1}{s + p_1} + \frac{R_2}{s + p_2} + \ldots + \frac{R_n}{s + p_n} \tag{7.31}$$

where $p_1, p_2, \ldots, p_n$ are the system poles and $R_1, R_2, \ldots, R_n$ the associated residues. Next, the following matrices can be written:

$$
A = \begin{bmatrix} -p_1 & & & \\ & -p_2 & & \\ & & \ddots & \\ & & & -p_n \end{bmatrix} ; \quad B = \begin{bmatrix} b_1 \\ b_2 \\ \vdots \\ b_n \end{bmatrix} ; \tag{7.32}
$$
$$
C = \begin{bmatrix} c_1 & c_2 & \cdots & c_n \end{bmatrix} ; \qquad D = 0
$$

with $c_1 b_1 = R_1$, $c_2 b_2 = R_2, \ldots$, and $c_n b_n = R_n$.

7.4 BIBO stability

Here we introduce the notion of Bounded-Input Bounded-Output (BIBO) stability.

A system is BIBO stable if for any bounded input, i.e., $|u(t)| < M_u$, the output is also bounded, i.e., $|y(t)| < M_y$, where M_u and M_y are two positive constants.

The property of BIBO stability is linked to the position of the poles in the complex plane: a system is BIBO stable if and only if all system poles have a negative real part.

BIBO stability and asymptotic stability are related to each other. In particular, if a system is asymptotically stable then it is also BIBO stable. Viceversa, a BIBO stable system may eventually be not asymptotically stable. Consider for instance the system described by the following state matrices:

$$
A = \begin{bmatrix} -1 & & \\ & 2 & \\ & & -5 \end{bmatrix} ; \quad B = \begin{bmatrix} 1 \\ 3 \\ 3 \end{bmatrix} ; \tag{7.33}
$$
$$
C = \begin{bmatrix} 1 & 0 & 1 \end{bmatrix} ; \qquad D = 0
$$

The transfer function of this system is

$$
G(s) = \frac{4(s+2)}{(s+1)(s+5)} \tag{7.34}
$$

The system is BIBO stable as the poles are $s_1 = -1$ and $s_2 = -5$. However, the system is not asymptotically stable, but unstable, because of the presence of $\lambda_2 = 2$.

In fact, asymptotic stability and BIBO stability coincide when the system is in minimal form, that is when system poles and system eigenvalues coincide.

We now consider a few examples. We describe the systems directly giving their transfer functions. In such cases, we should assume that there are no cancellations and that the system order is the order of the polynomial at the denominator.

Example 7.6 _____

The system $G(s) = \frac{s-3}{(s+4)(s+5)}$ is BIBO stable, and, for what is discussed above, also asymptotically stable, as all poles have a negative real part.

Example 7.7 _____

The system $G(s) = \frac{s-3}{(s-4)(s+5)}$ is unstable. It is easy to check that any bounded input yields an unbounded output, because of the presence of the mode e^{4t} associated with the positive pole.

Example 7.8 _____

The system $G(s) = \frac{1}{s}$ is not BIBO stable. To be BIBO stable, in fact, any bounded input should produce a bounded output. For this system, many bounded inputs lead to bounded outputs. However, the step input yields $y(t) = t$, which is clearly unbounded.

Example 7.9 _____

Also the system $G(s) = \frac{1}{s^2+2}$ is not BIBO stable. Similarly to the previous example, here, many bounded inputs lead to bounded outputs. The inputs that produce an unbounded output are of the type $u(t) = \sin(\sqrt{2}t)$. In such cases, the input produces a resonance in the system. We will come back to this phenomenon when studying the frequency response of a system.

As we have seen, BIBO stability depends on the pole location. The position of the zeros is, instead, important to define minimum phase systems. These are BIBO stable systems for which all zeros have a negative real part, that is, they are on the open left half-plane.

7.5 Zero-pole and Bode form of the transfer function

Two are the most used forms of the transfer function:

$$G(s) = \frac{\rho \Pi_i (s+z_i) \Pi_i (s^2 + 2\zeta_i \omega_{z,i} s + \omega_{z,i}^2)}{s^g \Pi_i (s+p_i) \Pi_i (s^2 + 2\xi_i \omega_{n,i} s + \omega_{n,i}^2)} \qquad (7.35)$$

and

$$G(s) = \frac{\mu \Pi_i (s/z_i + 1) \Pi_i (s^2/\omega_{z,i}^2 + 2\zeta_i s/\omega_{z,i} + 1)}{s^g \Pi_i (s/p_i + 1) \Pi_i (s^2/\omega_{n,i}^2 + 2\xi_i s/\omega_{n,i} + 1)} \qquad (7.36)$$

The first is called zero-pole form, the second Bode form. As it can be noticed, complex and conjugates poles are associated with second-order polynomials, in order to avoid the use of complex coefficients.

In these forms g represents the system type, z_i are the real zeros, p_i the system real poles, $\omega_{z,i}$ and $\omega_{n,i}$ the natural frequencies associated with complex and conjugates zeros and poles, ρ the transfer constant, and μ the system gain.

7.6 Final value theorem

There is an important property of the Laplace transform of a signal, known as the final value theorem, which is now discussed, as the notion of BIBO stability can help in understanding whether the result can be applied or not.

Given a generic signal $y(t)$, assumed that its limit for $t \to +\infty$ exists and is finite, then

$$\lim_{t \to +\infty} y(t) = \lim_{s \to 0} sY(s) \tag{7.37}$$

This result offers the possibility of calculating the final value of a signal (namely its steady-state value), if it exists and is finite, by calculating a limit in the s domain.

Let us now apply this theorem to LTI systems.

More specifically, let us consider a BIBO stable system written in Bode form (7.36) and assume that $g = 0$. Since the system is BIBO stable, then all poles have negative real part. Let us now consider the response of such a system to a step of amplitude $\bar{u}$. Because of the BIBO stability of the system, the limit of the signal for $t \to +\infty$ exists (in fact, the step response is given by the linear combination of the system modes and step(t)). Applying the final value theorem we obtain that

$$\bar{y} = \lim_{t \to +\infty} y(t) = \lim_{s \to 0} sY(s) = \mu\bar{u} \tag{7.38}$$

Therefore, μ represents the ratio between the steady-state value reached by the output, namely $\bar{y}$, and the amplitude of the input, $\bar{u}$.

Notice that g cannot be positive, as the system is BIBO stable. If $g < 0$, then $\bar{y} = 0$, because of the presence of one or more zeros at the origin.

7.7 Systems with dead time

Time-delayed systems or systems with dead time are characterized by a transfer function of this type

$$G(s) = \frac{N(s)}{D(s)} e^{-sT} \tag{7.39}$$

where T is the dead time. This is what occurs in the presence of a time-delayed input $u(t - T)$, due for instance to some transport phenomenon present in the system. Notice that the presence of a delay introduces a non-rational term e^{-sT}, which can significantly impact the stability and performance when the system operates in a feedback loop.

An example of such systems is a first-order system with dead-time, which can be modeled as follows:

$$G(s) = \frac{\mu}{\tau s + 1} e^{-sT} \tag{7.40}$$

where μ is the static gain, and τ the time constant of the system.

The delay term can be dealt with using the Padé approximation

$$e^{-sT} \simeq \frac{1 - \frac{T}{2}s}{1 + \frac{T}{2}s} \tag{7.41}$$

Example 7.10 _____

Consider the pendulum whose transfer function was derived in Example 7.1. Suppose that the torque applied to the pendulum is subject to a time delay. Under this assumption, the transfer function of the system becomes:

$$G(s) = \frac{\frac{1}{ml^2}}{s^2 + \frac{g}{l}} e^{-sT} \tag{7.42}$$

Using the Padé approximation we obtain

$$G(s) = \frac{\frac{1}{ml^2}}{s^2 + \frac{g}{l}} \frac{2 - Ts}{2 + Ts} \tag{7.43}$$

Notice the presence of a positive zero.

7.8 Calculation of the transfer function in MATLAB

In this section we illustrate how to calculate the transfer function of an LTI system in MATLAB®. To illustrate the procedure, we refer to the quarter-car model of Sec. 7.3. We will do the calculation with two methods. In the first case, we will fix the system parameters and, then, calculate the transfer function. In the second case, we will carry out a symbolic calculation of the transfer function.

In the first case, we first define the parameters and the state matrices as follows:

```
M1=1;
M2=10;
ks=2;
kw=3;
b=0.5;
A=[0 0 1 0;
   0 0 0 1;
   -(ks+kw)/M1 ks/M1 -b/M1 b/M1;
   ks/M2 -ks/M2 b/M2 -b/M2];
B=[0; 0; kw/M1; 0];
C=[0 1 0 0];
D=0;
quartercarmodel=ss(A,B,C,D)
```

Then, we use the command **tf** that computes the transfer function given as input to the LTI object defined above with the command **ss**:

```
tf(quartercarmodel)
```

Note that the command **tf** can be used also to define a system model starting from the knowledge of its transfer function.

In the second approach, we compute the transfer function of the model using the symbolic toolbox of MATLAB®. To this aim, we first define the parameters as symbolic variables, specifying that they are real values:

```
syms M1 M2 b ks kw real
```

In a similar fashion we define the symbolic variable s. Since the variable is complex and the symbolic toolbox by default assumes that variables are such, we can use the following command:

```
syms s
```

Once we have defined the parameters and the complex variable s, we define the matrices, which will be symbolic variables as well:

```
A=[0 0 1 0;
   0 0 0 1;
   -(ks+kw)/M1 ks/M1 -b/M1 b/M1;
   ks/M2 -ks/M2 b/M2 -b/M2];
B=[0; 0; kw/M1; 0];
C=[0 1 0 0];
D=0;
```

Since the commands **ss** and **tf** work for numerical computation, here to calculate the transfer function we apply Eq. (7.4) as follows:

```
G=C*inv(s*eye(4)-A)*B+D
```

Transfer functions of time-delayed systems can be defined as well. Consider, for instance, the first-order system with dead time $G(s) = \frac{2}{5s+1}e^{-3s}$, where the delay is $T = 3$. To define the system, these commands can be used:

```
s=tf('s')
G=2/(5*s+1)*exp(-3*s)
```

Padé approximations can also be used for approximating the delay term. The command to use is **pade(T,N)**, which requires as inputs the time delay T and the order N of the approximation. For example, the commands

```
T=3
[num, den]=pade(T,1)
```

produce the approximation as in Eq. (7.41).

7.9 Exercises

1. Calculate the transfer function for the system with the following state matrices:

$$A = \begin{bmatrix} -3 & 1 \\ 1 & 2 \end{bmatrix}; \quad B = \begin{bmatrix} 1 \\ 0 \end{bmatrix}; \tag{7.44}$$
$$C = \begin{bmatrix} 0 & 2 \end{bmatrix}; \quad D = 2$$

2. Calculate the transfer function for the system with the following state matrices:

$$A = \begin{bmatrix} -3 & 0 \\ 0 & -4 \end{bmatrix}; \quad B = \begin{bmatrix} 1 \\ 2 \end{bmatrix}; \tag{7.45}$$
$$C = \begin{bmatrix} 1.5 & 3 \end{bmatrix}; \quad D = 0$$

3. Given a system with transfer function $G(s) = \frac{s+3}{s^4+1.5s^3+0.3s^2+2s+3.3}$ determine whether it is BIBO stable or not.

4. Given a system with transfer function $G(s) = \frac{5}{s^3+0.1s^2+2s+0.2}$ determine whether it is BIBO stable or not.

5. For the system with transfer function $G(s) = \frac{s^2+s+3}{s^4+13s^2+36}$, find an example of a bounded input that produces an unbounded output.

8

Lecture 8—System aggregates

In this lecture we study aggregates of systems. Namely, we study how two systems may interact and the equivalent transfer function that is obtained from this interaction. We illustrate three configurations: series/cascade, parallel, and feedback configuration.

8.1   Series/cascade connection

In the series, or cascade, connection the output of a system is the input of a second system, as in Fig. 8.1. Here, we consider the more general case of interconnected MIMO systems and derive the equivalent transfer matrix $G(s)$ as follows:

$$\mathbf{Y}(s) = \mathbf{Y}_2(s) = G_2(s)\mathbf{U}_2(s) \tag{8.1}$$

Since $\mathbf{U}_2(s) = \mathbf{Y}_1(s)$, then

$$\mathbf{Y}(s) = G_2(s)\mathbf{Y}_1(s) = G_2(s)G_1(s)\mathbf{U}(s) \tag{8.2}$$

We conclude that

$$G(s) = G_2(s)G_1(s) \tag{8.3}$$

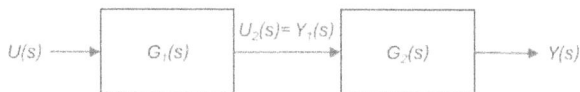

FIGURE 8.1
Series/cascade configuration.

DOI: 10.1201/9781003487289-8

8.2 Parallel connection

In the parallel configuration (Fig. 8.2) the same input $u(t)$ is applied to two systems, whose outputs are then summed. To find the equivalent transfer matrix $G(s)$, we consider that

$$\mathbf{Y}(s) = \mathbf{Y}_1(s) + \mathbf{Y}_2(s) \tag{8.4}$$

Then, since $\mathbf{Y}_1(s) = G_1(s)\mathbf{U}_1(s)$ and $\mathbf{Y}_2(s) = G_2(s)\mathbf{U}_2(s)$, we get:

$$\mathbf{Y}(s) = G_1(s)\mathbf{U}_1(s) + G_2(s)\mathbf{U}_2(s) = (G_1(s) + G_2(s))\mathbf{U}(s) \tag{8.5}$$

We conclude that for the parallel connection the equivalent transfer function is given by

$$G(s) = G_1(s) + G_2(s) \tag{8.6}$$

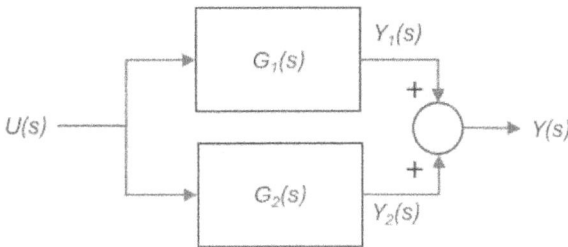

FIGURE 8.2
Parallel configuration.

8.3 Feedback connection

In the feedback configuration (Fig. 8.3), the output $\mathbf{Y}(s)$ of the system with transfer matrix $G_1(s)$ is fed back, through the system with transfer matrix $G_2(s)$, to the input stage, where it is compared with the input $\mathbf{U}(s)$.

To find the equivalent transfer matrix, we write the system output as:

$$\mathbf{Y}(s) = G_1(s)(\mathbf{U}(s) - G_2(s)\mathbf{Y}(s)) \tag{8.7}$$

From this expression we derive that

$$\mathbf{Y}(s) + \mathbf{G}_1(s)\mathbf{G}_2(s)\mathbf{Y}(s) = \mathbf{G}_1(s)\mathbf{U}(s) \tag{8.8}$$

and hence

$$\mathbf{Y}(s) = (\mathbf{I} + \mathbf{G}_1(s)\mathbf{G}_2(s))^{-1}\mathbf{G}_1(s)\mathbf{U}(s) \tag{8.9}$$

Summing up, the equivalent transfer matrix of the feedback configuration for MIMO systems is

$$\mathbf{G}(s) = (\mathbf{I} + \mathbf{G}_1(s)\mathbf{G}_2(s))^{-1}\mathbf{G}_1(s) \tag{8.10}$$

and for SISO systems

$$G(s) = \frac{G_1(s)}{1 + G_1(s)G_2(s)} \tag{8.11}$$

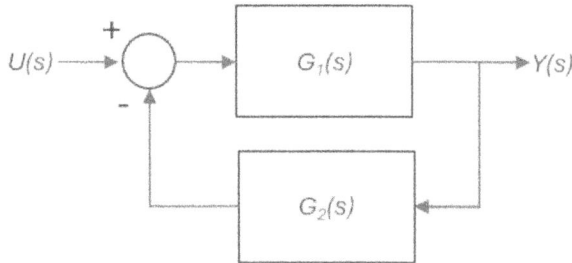

FIGURE 8.3
Feedback configuration.

In case of positive feedback, for MIMO systems the transfer matrix reads:

$$\mathbf{G}(s) = (\mathbf{I} - \mathbf{G}_1(s)\mathbf{G}_2(s))^{-1}\mathbf{G}_1(s) \tag{8.12}$$

and for SISO systems the transfer function reads:

$$G(s) = \frac{G_1(s)}{1 - G_1(s)G_2(s)} \tag{8.13}$$

8.4   System aggregates and BIBO stability

Let $G_1(s) = \frac{N_1(s)}{D_1(s)}$ and $G_2(s) = \frac{N_2(s)}{D_2(s)}$, then for the series connection we have that

$$G(s) = \frac{N_1(s)N_2(s)}{D_1(s)D_2(s)} \tag{8.14}$$

In the absence of cancellations between the roots of $N_1(s)N_2(s)$ and the roots of $D_1(s)D_2(s)$, the poles of the system are the union of those of the two subparts. This means that, if $G_1(s)$ and $G_2(s)$ are BIBO stable, so is their series connection.

For the parallel configuration, we have that

$$G(s) = \frac{N_1(s)D_2(s) + N_2(s)D_1(s)}{D_1(s)D_2(s)} \tag{8.15}$$

Hence, in the absence of cancellations between the roots of $N_1(s)D_2(s) + N_2(s)D_1(s)$ and the roots of $D_1(s)D_2(s)$, the poles of the system are the union of those of the two subparts. Also, in this case, two BIBO stable systems connected in parallel yield a BIBO stable system.

Summing up, we have that the series or the parallel connection of two BIBO stable systems always yield a system that is BIBO stable. If one of the two systems is not BIBO stable, then also the series or the parallel connection is not BIBO stable.

Finally, for the feedback configuration, we have that

$$G(s) = \frac{N_1(s)D_2(s)}{N_1(s)N_2(s) + D_1(s)D_2(s)} \tag{8.16}$$

In this case the relationship between the stability of the two interacting systems and that of the closed-loop system is not trivial. Two BIBO stable systems in feedback configurations can eventually yield a system that is not stable. Conversely, the closed-loop system can be BIBO stable even if $G_1(s)$ and $G_2(s)$ are not BIBO stable.

8.5   Moving blocks

Figs. 8.4 and 8.5 illustrate how to move blocks in a block diagram. The rule is that the same output signal should be obtained before and after moving the block. For instance, from Fig. 8.4(a) we obtain that $Y(s) = G(s)(U_1(s) + U_2(s))$. From the equivalent diagram in Fig. 8.4(b), we also obtain that $Y(s) = G(s)(U_1(s) + U_2(s))$.

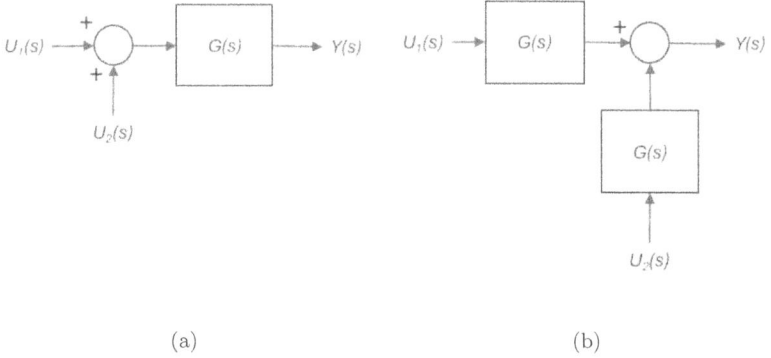

(a)   (b)

FIGURE 8.4
Shifting a block into a position before a summing point.

Similarly, in both diagrams of Fig. 8.5 the output is given by: $Y(s) = G(s)U_1(s) + U_2(s)$.

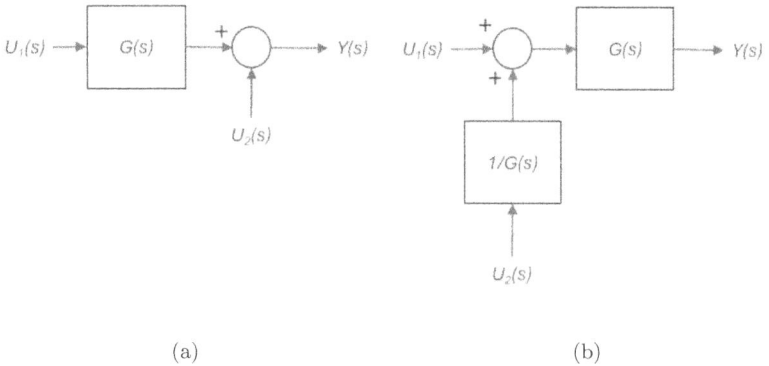

(a)   (b)

FIGURE 8.5
Shifting a block into a position after a summing point.

8.6 Block diagram of the state-space equations

Fig. 8.6 shows the block diagram of the state-space equations of an LTI system. The output of the first summing block is given by $\mathbf{Bu} + \mathbf{Ax}$, that is $\dot{\mathbf{x}}$, the derivative of the state variable. This signal is the input of an integral block, such that the output is the state vector. In turn the state vector is multiplied by $\mathbf{C}$ and added to the signal $\mathbf{Du}$ to form the output $\mathbf{y}$ of the system. In this way, the block diagram of Fig. 8.6 is exactly equivalent to the state space equations (2.2) of a system.

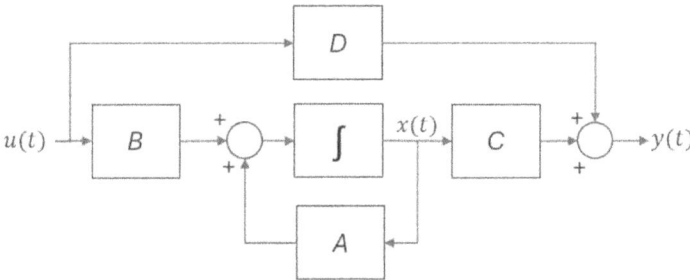

FIGURE 8.6
Block diagram of the state space equations.

8.7 Example of block diagram reduction

We illustrate now an example of how to find the equivalent transfer function of a block diagram constituted by several interacting subparts.

Example 8.1 _____

Let us consider the block diagram shown in Fig. 8.7.
To calculate the equivalent transfer function of the diagram, we proceed with successive iterations, where at each iteration we identify one of the three elementary configurations previously discussed, substitute the equivalent transfer function of the identified config- uration, and then redraw the diagram. These steps are shown in Fig. 8.8. In the first step (Fig. 8.8(a)), we identify the parallel connection of blocks $G_1(s)$ and $G_2(s)$. We replace this configuration with a single transfer function $G_1(s) + G_2(s)$ and redraw the block diagram as in Fig. 8.8(b). In this new diagram we identify the feedback configuration made by the block $G_1(s) + G_2(s)$ and the block $G_4(s)$. The equivalent transfer function is $\frac{G_1(s) + G_2(s)}{1 + (G_1(s) + G_2(s))G_4(s)}$. We replace the feedback configuration with a single block with this transfer function and redraw again the block diagram as in Fig. 8.8(c). Next, we identify the series of two blocks, one with transfer function $\frac{G_1(s) + G_2(s)}{1 + (G_1(s) + G_2(s))G_4(s)}$ and the other with transfer function $G_3(s)$, such that the equivalent transfer function is the

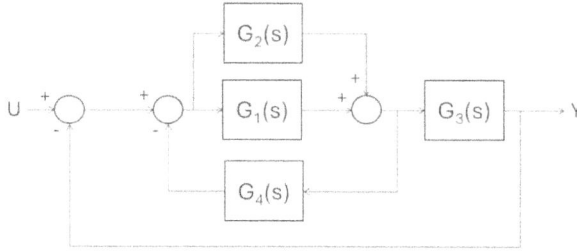

FIGURE 8.7
Example of a block scheme made of several interacting systems.

product of the two terms. Finally, in Fig. 8.8(d) we identify the feedback configuration made by the block $\frac{G_3(s)(G_1(s)+G_2(s))}{1+(G_1(s)+G_2(s))G_4(s)}$ and a block with unit transfer function. Taking into account this last step, we conclude that the transfer function of the block diagram of Fig. 8.7 is given by

$$G(s) = \frac{G_1(s)G_3(s) + G_2(s)G_3(s)}{1 + G_1(s)G_4(s) + G_2(s)G_4(s) + G_1(s)G_3(s) + G_2(s)G_3(s)} \qquad (8.17)$$

8.8   Stability of a block diagram as a function of one parameter

Now, let us study the stability of block diagrams containing a parameter. We will examine this through two examples.

Example 8.2 _____

Consider the block diagram of Fig. 8.9, where one of the blocks has a transfer function that depends on a parameter $k \in \mathbb{R}$. We want to study the stability of the system when this parameter is varied. To this aim, we first derive the equivalent transfer function of the diagram, and then apply the Routh criterion to the denominator of this transfer function.
In the block diagram shown in Fig. 8.9, we identify two elementary configurations, namely a cascade connection between the block $1/s^2$ and the one $1/(s + k)$, and a parallel connection between the block 2 and the block s. Taking into account these connections, the block diagram of Fig. 8.9 can be redrawn as in Fig. 8.10.
The block diagram of Fig. 8.10 consists of a negative feedback connection of the block with transfer function $1/(s^2(s + k))$ and the one $s + 2$, such that the overall transfer function reads:

$$G(s) = \frac{1}{s^3 + ks^2 + s + 2} \qquad (8.18)$$

The stability of the system is studied by applying the Routh criterion to the polynomial $p(s) = s^3 + ks^2 + s + 2$. The corresponding Routh table is Table 8.1. From this table, we conclude that for $k > 2$ there are three poles with a negative real part, so the closed-loop system is BIBO stable; for $k = 2$ there are two poles with zero real part, and one

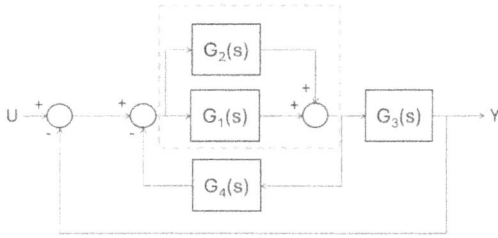

(a)

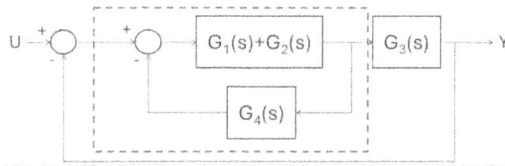

(b)

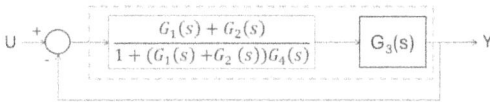

(c)

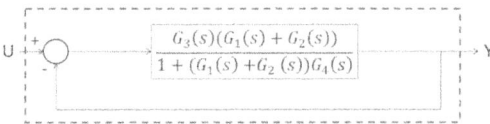

(d)

FIGURE 8.8
Steps to find the equivalent transfer function of the block diagram of Fig. 8.7.

with a negative real part, so that the system is stable but not BIBO stable; for $k < 2$ there are two poles with positive real part, and one with negative real part, so that the system is unstable.

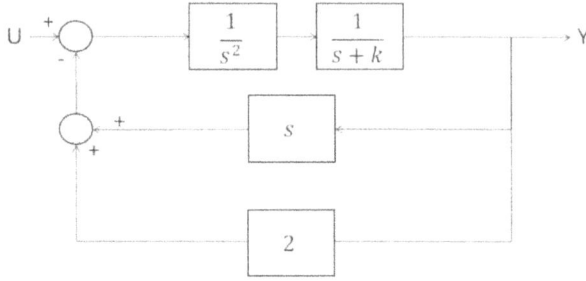

FIGURE 8.9
Example of a block diagram made of several interacting systems, one of which contains a parameter k.

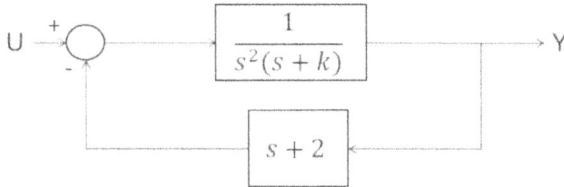

FIGURE 8.10
Block diagram equivalent to the one in Fig. 8.9.

Example 8.3 _____

Consider now the block diagram in Fig. 8.11.
The closed-loop transfer function is given by

$$F(s) = \frac{1}{(s+1)^\rho + k} \tag{8.19}$$

For $\rho = 1$ we have

$$F(s) = \frac{1}{s+1+k} \tag{8.20}$$

which is stable if $k > -1$.
For $\rho = 2$ we have

$$F(s) = \frac{k}{s^2 + 2s + 1 + k} \tag{8.21}$$

which is stable if $k > -1$.
For $\rho = 3$ we have

$$F(s) = \frac{1}{s^3 + 3s^2 + 3s + 1 + k} \tag{8.22}$$

TABLE 8.1

Routh table for $p(s) = s^3 + ks^2 + s + 2$.

3	1	1
2	k	2
1	$\frac{k-2}{k}$	
0	2	

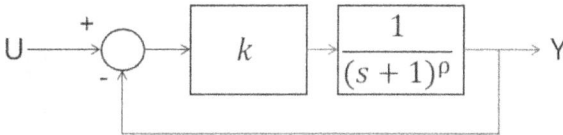

FIGURE 8.11

Example of a block diagram which contains a parameter k.

The corresponding Routh table (Table 8.2) indicates that the system is stable if $-1 < k < 8$.

For $\rho = 4$ we have

$$F(s) = \frac{1}{s^4 + 4s^3 + 6s^2 + 4s + 1 + k} \tag{8.23}$$

The corresponding Routh table (Table 8.3) indicates that the system is stable if $-1 < k < 4$.

Taken together, the results suggest that as the system order ρ increases, the range of k values that ensure closed-loop stability decreases.

8.9 MATLAB commands for the study of system aggregates

MATLAB® provides three commands for the corresponding configurations of two interacting systems: series, parallel, and feedback. As an example of their use, we consider the block diagram of Fig. 8.7 and find the transfer function of the equivalent diagram.

TABLE 8.2

Routh table for the polynomial $p(s) = s^3 + 3s^2 + 3s + 1 + k$.

3	1	3
2	3	1+k
1	$\frac{8-k}{3}$	
0	1+k	

TABLE 8.3

Routh table for the polynomial $p(s) = s^4 + 4s^3 + 6s^2 + 4s + 1 + k$.

4	1	6	1+k
3	4	4	
2	5	1+k	
1	$\frac{16-4k}{5}$		
0	1+k		

First, we define the four subsystems involved in the diagram, namely $G_1(s)$, $G_2(s)$, $G_3(s)$, and $G_4(s)$.

```
s=tf('s')
k=5;
G1=1/s^2;
G2=1/(s+k);
G3=s;
G4=2;
```

Next, we build the block diagram of Fig. 8.7, taking into account that $G_1(s)$ and $G_2(s)$ are in series (we indicate the resulting, equivalent system as $G_5(s)$), that $G_3(s)$, and $G_4(s)$ are in parallel (we indicate the resulting, equivalent system as $G_6(s)$), and, lastly, that $G_5(s)$ and $G_6(s)$ are in feedback configuration, with $G_5(s)$ in the direct chain, and $G_6(s)$ in the feedback chain. We can, thus, use the following commands:

```
G5=series(G1,G2)
G6=parallel(G3,G4)
G=feedback(G5,G6)
```

The transfer function of the equivalent system fully describes the system behavior, and we can use it, for instance, to calculate the closed-loop step response, or map the system poles and zeros, as follows:

```
figure,step(G)
figure,pzmap(G)
```

8.10   Exercises

1. Derive the transfer function of the block diagram of Fig. 8.12 and study the stability with respect to the parameter k.

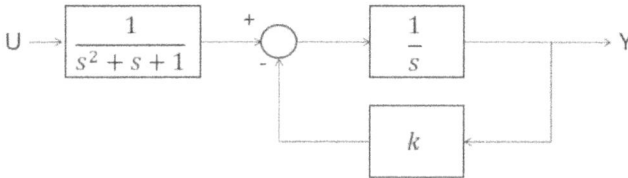

FIGURE 8.12
Block diagram for Exercise 1.

2. Derive the transfer function of the block diagram of Fig. 8.13 and study the stability with respect to the parameter k.

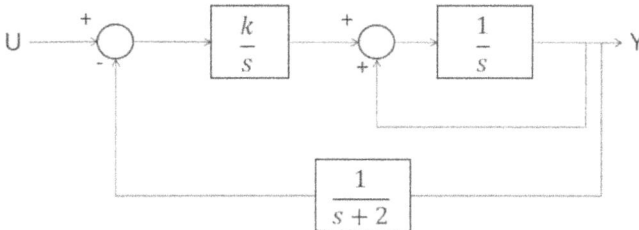

FIGURE 8.13
Block diagram for Exercise 2.

3. Derive the transfer function of the block diagram of Fig. 8.14 and study the stability with respect to the parameter k.

4. Derive the transfer function of the block diagram of Fig. 8.15 and study the stability with respect to the parameter k.

5. Calculate the step response for the system represented in Fig. 8.16.

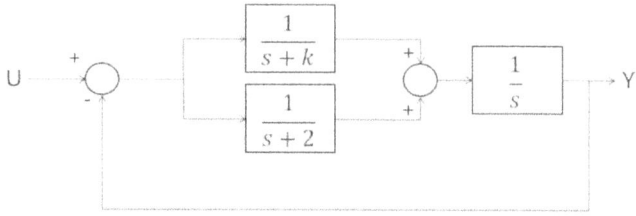

FIGURE 8.14
Block diagram for Exercise 3.

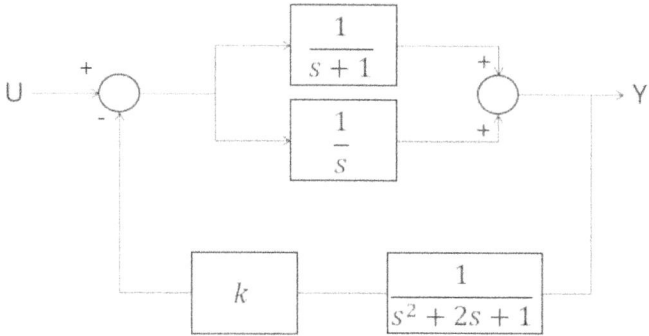

FIGURE 8.15
Block diagram for Exercise 4.

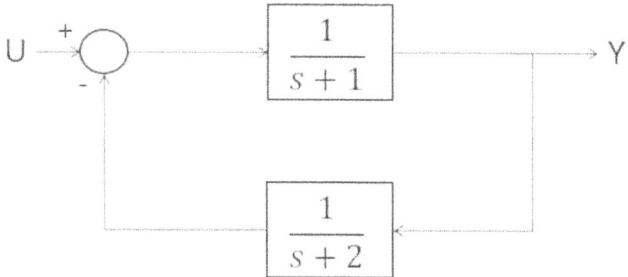

FIGURE 8.16
Block diagram for Exercise 5.

9

Lecture 9—Reachability and observability of continuous time linear systems

In this lecture, we study two important properties of linear systems, namely reachability and observability. With reference to continuous time systems, we begin defining reachable states and reachable systems, and introduce the reachability canonical form. Then, we study the problem of observability, introducing the notion of observable states and observable systems, as well as the observability canonical form. We end the chapter by studying the MATLAB® commands related to this topic.

9.1 Reachability and controllability

We consider continuous-time LTI systems governed by the equations

$$\begin{aligned} \dot{\mathbf{x}} &= \mathbf{Ax} + \mathbf{Bu} \\ \mathbf{y} &= \mathbf{Cx} + \mathbf{Du} \end{aligned} \tag{9.1}$$

A state $\tilde{\mathbf{x}}$ is said to be reachable if there exist $\tilde{t} > 0$ and $\tilde{\mathbf{u}}(t)$ with $t \in [0, \tilde{t}]$ such that the system evolution starting from $\mathbf{x}_0$ reaches the state $\tilde{\mathbf{x}}$ at time $\tilde{t}$, i.e., $\mathbf{x}(\tilde{t}) = \tilde{\mathbf{x}}$. The problem considered is whether it is possible to find an input such that the system states can be moved in a finite time $[0, \tilde{t}]$ from the initial state $\mathbf{x}_0 = 0$ to the final state $\tilde{\mathbf{x}}$. A system is said completely reachable if all states are reachable.

Conversely, a state $\tilde{\mathbf{x}}$ is said to be controllable if there exist $\tilde{t} > 0$ and $\tilde{\mathbf{u}}(t)$ with $t \in [0, \tilde{t}]$ such that the system starting from $\tilde{\mathbf{x}}$ can be moved into the state $\mathbf{x}_0 = 0$ at time $\tilde{t}$, i.e., $\mathbf{x}(\tilde{t}) = \mathbf{x}_0$. A system is completely controllable if all states are controllable.

Reachability and controllability describe the ability to modify a system's state by applying an appropriate input. For continuous-time LTI systems, reachability and controllability are indistinguishable properties. However, we will demonstrate that, for discrete-time systems, there can be states that are reachable but not controllable.

Reachability is a property that depends solely on A and B, as it is evident that the equation $\mathbf{y} = \mathbf{Cx} + \mathbf{Du}$ is not involved. To determine whether a system

DOI: 10.1201/9781003487289-9

is completely reachable, we use the reachability matrix, which is defined as follows:

$$M_r = \begin{bmatrix} B & AB & \cdots & A^{n-1}B \end{bmatrix} \tag{9.2}$$

A system is completely reachable if and only if the rank of the matrix is maximum, i.e., $\rho(M_r) = n$. For SISO systems, M_r is a square matrix and the condition for complete reachability simplifies to $\det(M_r) \neq 0$.

If the rank of M_r is less than n, it is possible to isolate the part of the system which satisfies the reachability property. Through the state transformation $\hat{x} = T_r^{-1}x$, Eqs. (9.1) can be rewritten as:

$$\begin{aligned} \dot{\hat{x}} &= \hat{A}\hat{x} + \hat{B}u \\ y &= \hat{C}\hat{x} + \hat{D}u \end{aligned} \tag{9.3}$$

with

$$\hat{A} = \begin{bmatrix} \hat{A}_a & \hat{A}_{ab} \\ 0 & \hat{A}_b \end{bmatrix} ; \hat{B} = \begin{bmatrix} \hat{B}_a \\ 0 \end{bmatrix} \tag{9.4}$$

Here, the remarkable property is the structure that the matrices $\hat{A}$ and $\hat{B}$ have, while the matrices $\hat{C}$ and $\hat{D}$ have no special structure. Specifically, the form of $\hat{A}$ and $\hat{B}$ allows one to rewrite Eqs. (9.3) as follows:

$$\begin{aligned} \dot{\hat{x}}_a &= \hat{A}_a\hat{x}_a + \hat{A}_{ab}\hat{x}_b + \hat{B}_a u \\ \dot{\hat{x}}_b &= \hat{A}_b\hat{x}_b \\ y &= \hat{C}\hat{x} + \hat{D}u \end{aligned} \tag{9.5}$$

This highlights how the system is divided into two parts: part a, which is reachable, is influenced by the state variables of part b, while part b, which is not reachable, is unaffected by either the state variables of part a or the input u.

The matrix T_r can be obtained by taking the first columns as the linearly independent columns of M_r and, then, selecting the remaining ones to have a square matrix in such a way that the matrix is invertible, i.e., $\det(T_r) \neq 0$. Therefore, the transformation $\hat{x} = T_r^{-1}x$ is not unique.

Example 9.1 _____

Consider the continuous-time LTI system with state matrices

$$A = \begin{bmatrix} 0 & 1 & 0 \\ -1 & 1 & 0 \\ 2 & -1 & -1 \end{bmatrix} ; \quad B = \begin{bmatrix} 0 \\ 1 \\ -1 \end{bmatrix} ; \tag{9.6}$$
$$C = \begin{bmatrix} 0 & 0 & 1 \end{bmatrix} ; \qquad D = 0$$

The reachability matrix of this system is

$$M_r = \begin{bmatrix} B & AB & A^2B \end{bmatrix} = \begin{bmatrix} 0 & 1 & 1 \\ 1 & 1 & 0 \\ -1 & 0 & 1 \end{bmatrix} \tag{9.7}$$

The rank of this matrix is less than the system order, as $\rho(M_r) = 2$, hence the system is not completely reachable.

To construct the matrix T_r, we start with the first two columns of M_r, which are linearly independent, and then add a third column chosen to be linearly independent from the first two. We get

$$T_r = \begin{bmatrix} 0 & 1 & 0 \\ 1 & 1 & 0 \\ -1 & 0 & 1 \end{bmatrix} \tag{9.8}$$

This matrix is invertible by construction. Using Eqs. (5.7), we obtain the state matrices in the new state space representation:

$$\hat{A} = T_r^{-1} A T_r = \begin{bmatrix} 0 & -1 & 0 \\ 1 & 1 & 0 \\ 0 & 0 & -1 \end{bmatrix} ; \quad \hat{B} = T_r^{-1} B = \begin{bmatrix} 1 \\ 0 \\ 0 \end{bmatrix} ;$$

$$\hat{C} = C T_r = \begin{bmatrix} -1 & 0 & 1 \end{bmatrix} ; \quad D = 0 \tag{9.9}$$

Eqs. (9.9) are in the form of Eqs. (9.5) with a reachable part (part a) of order two, and a non-reachable part (part b) of order one.

In the state-space representation in Eqs. (9.9), it is immediate to identify the states that are reachable. In fact, all states $\hat{\mathbf{x}} = \begin{bmatrix} \gamma_1 \\ \gamma_2 \\ 0 \end{bmatrix}$ with $\gamma_1, \gamma_2 \in \mathbb{R}$ are reachable. This

yields that, in the original state-space representations, the states $\mathbf{x} = T_r \begin{bmatrix} \gamma_1 \\ \gamma_2 \\ 0 \end{bmatrix} =$

$\begin{bmatrix} \gamma_2 \\ \gamma_1 + \gamma_2 \\ -\gamma_1 \end{bmatrix}$ are reachable.

A system that is completely reachable can be written in *reachability canonical form*. This state-space representation is defined by a special form for the matrices A and B, as follows:

$$A^{(r)} = \begin{bmatrix} 0 & 1 & 0 & \cdots & 0 \\ 0 & 0 & 1 & \cdots & 0 \\ \vdots & & & & \vdots \\ 0 & 0 & 0 & \cdots & 1 \\ -\alpha_n & -\alpha_{n-1} & -\alpha_{n-2} & \cdots & -\alpha_1 \end{bmatrix} ; \quad B^{(r)} = \begin{bmatrix} 0 \\ 0 \\ \vdots \\ 0 \\ 1 \end{bmatrix} ;$$

$$C^{(r)} = \begin{bmatrix} b_n & b_{n-1} & b_{n-2} & \cdots & b_1 \end{bmatrix} ; \quad D^{(r)} = d \tag{9.10}$$

where $\alpha_1, \alpha_2, \ldots, \alpha_n$ and $b_1, b_2, \ldots, b_n$ are the coefficients of the polynomials in the denominator and numerator of the transfer function, respectively, that is

$$G(s) = \frac{b_1 s^{n-1} + \ldots + s b_{n-1} + b_n}{s^n + \alpha_1 s^{n-1} + \ldots + \alpha_n} + d \tag{9.11}$$

The reachability canonical form can be derived by applying the state transformation $\tilde{\mathbf{x}} = T^{-1}\mathbf{x}$ where T is constructed based on the relationship between M_r and $M_r^{(r)}$. In fact, taking into account Eqs. (5.7), we find that

$$
\begin{aligned}
M_r^{(r)} &= \begin{bmatrix} B^{(r)} & A^{(r)}B^{(r)} & \cdots & (A^{(r)})^{n-1}B \end{bmatrix} \\
&= \begin{bmatrix} T^{-1}B & T^{-1}ATT^{-1}B & \cdots & T^{-1}A^{n-1}TT^{-1}B \end{bmatrix} \\
&= T^{-1}\begin{bmatrix} B & AB & \cdots & A^{n-1}B \end{bmatrix} \\
&= T^{-1}M_r
\end{aligned}
\tag{9.12}
$$

and so

$$
T = M_r(M_r^{(r)})^{-1}
\tag{9.13}
$$

Example 9.2 _____

Consider the continuous-time LTI system with state matrices:

$$
A = \begin{bmatrix} 1 & -1 & 3 \\ -2 & 0 & 4 \\ 1 & 1 & -3 \end{bmatrix}; \quad B = \begin{bmatrix} 2 \\ -1 \\ 1 \end{bmatrix};
$$
$$
C = \begin{bmatrix} 0 & 1 & 1 \end{bmatrix}; \quad D = 0
\tag{9.14}
$$

The reachability matrix of this system is

$$
M_r = \begin{bmatrix} B & AB & A^2B \end{bmatrix} = \begin{bmatrix} 2 & 6 & 0 \\ -1 & 0 & -20 \\ 1 & -2 & 12 \end{bmatrix}
\tag{9.15}
$$

Since the rank of the matrix is three, the system is completely reachable, and we can express an equivalent state-space representation in the reachability canonical form. To do this, we first compute the transfer function:

$$
G(s) = C(sI - A)^{-1}B + D = \frac{-2s - 12}{s^3 + 2s^2 - 12s + 8}
\tag{9.16}
$$

This allows us to obtain the coefficients of the numerator and denominator polynomials needed to write the state matrices of the reachability canonical form:

$$
A^{(r)} = \begin{bmatrix} 0 & 1 & 0 \\ 0 & 0 & 1 \\ -8 & 12 & -2 \end{bmatrix}; \quad B^{(r)} = \begin{bmatrix} 0 \\ 0 \\ 1 \end{bmatrix};
$$
$$
C^{(r)} = \begin{bmatrix} -12 & -2 & 0 \end{bmatrix}; \quad D^{(r)} = 0
\tag{9.17}
$$

9.2   Observability

In this section, we discuss the concept of observability for continuous-time LTI systems. We start with the definition of states that are not observable.

A state $\tilde{\mathbf{x}} \neq 0$ is said to be unobservable if, indicated with $\tilde{\mathbf{y}}_n(t)$ the natural response of the system starting from $\tilde{\mathbf{x}}$, then, for any $\tilde{t} > 0$ we have that $\tilde{\mathbf{y}}_n(t) = 0$ for $t \in [0, \tilde{t}]$. A system with no unobservable states is said to be completely observable.

Since the natural response is involved, the observability property does not depend on the input and, consequently, on the matrix B. The complete observability of a system can be determined using the observability matrix, which is defined as follows:

$$M_o = \begin{bmatrix} C \\ CA \\ \vdots \\ CA^{n-1} \end{bmatrix} \tag{9.18}$$

A system is completely observable if and only if the rank of the matrix M_o is maximum, i.e., $\rho(M_o) = n$. For SISO systems, M_o is a square matrix and the condition for complete observability is that $\det(M_o) \neq 0$.

If $\rho(M_o) < n$, then the system can be divided into two parts: one that is unobservable, and another that is observable. This can be achieved through the state transformation $\hat{\mathbf{x}} = T_o^{-1}\mathbf{x}$ that allows us to rewrite Eqs. (9.1) as follows:

$$\begin{aligned} \dot{\hat{\mathbf{x}}} &= \hat{A}\hat{\mathbf{x}} + \hat{B}\mathbf{u} \\ \mathbf{y} &= \hat{C}\hat{\mathbf{x}} + \hat{D}\mathbf{u} \end{aligned} \tag{9.19}$$

with

$$\begin{aligned} \hat{A} &= \begin{bmatrix} \hat{A}_a & 0 \\ \hat{A}_{ba} & \hat{A}_b \end{bmatrix} \\ \hat{C} &= \begin{bmatrix} \hat{C}_a & 0 \end{bmatrix} \end{aligned} \tag{9.20}$$

In this case, the matrices $\hat{A}$ and $\hat{C}$ have a peculiar block structure, while the matrices $\hat{B}$ and $\hat{D}$ have no special structure. Taking into account the structure of $\hat{A}$ and $\hat{C}$, Eqs. (9.19) can be rewritten as follows:

$$\begin{aligned} \dot{\hat{\mathbf{x}}}_a &= \hat{A}_a\hat{\mathbf{x}}_a + \hat{B}_a\mathbf{u} \\ \dot{\hat{\mathbf{x}}}_b &= \hat{A}_{ba}\hat{\mathbf{x}}_a + \hat{A}_b\hat{\mathbf{x}}_b + \hat{B}_b\mathbf{u} \\ \mathbf{y} &= \hat{C}_a\hat{\mathbf{x}}_a + \hat{D}\mathbf{u} \end{aligned} \tag{9.21}$$

Eqs. (9.21) show that the system is divided into two parts: part a, which is observable, and part b, which is unoservable. Notice that the output $\mathbf{y}$ is not influenced by the variables of the unoservable part, neither directly, as $\hat{\mathbf{x}}_b$ does not appear in the equation of $\mathbf{y}$ or indirectly, as $\hat{\mathbf{x}}_b$ does not affect the dynamics of the variables of the observable part.

Similarly to what was observed for reachability, the transformation $\hat{\mathbf{x}} = T_o^{-1}\mathbf{x}$ is not unique. The matrix T_o can be constructed as follows: first, T_o^{-1} is built by selecting its first rows as the linearly independent rows of M_o and, then, completing it with additional (linearly independent) rows to form a square, invertible matrix, i.e., $\det(T_o^{-1}) \neq 0$. Finally, T_o is obtained as the inverse of T_o^{-1}.

Example 9.3 _____

Consider the continuous-time LTI system with state matrices:

$$A = \begin{bmatrix} 0 & -1 & 1 \\ 1 & 0 & 1 \\ 1 & -1 & 0 \end{bmatrix}; \quad B = \begin{bmatrix} 1 \\ -1 \\ 2 \end{bmatrix}; \quad (9.22)$$
$$C = \begin{bmatrix} 1 & 0 & 1 \end{bmatrix}; \quad D = 0$$

The observability matrix of this system is

$$M_o = \begin{bmatrix} C \\ CA \\ \vdots \\ CA^{n-1} \end{bmatrix} = \begin{bmatrix} 1 & 0 & 1 \\ 1 & -2 & 1 \\ -1 & -2 & -1 \end{bmatrix} \quad (9.23)$$

Since $\rho(M_o) = 2$, the system is not completely observable.
Next, the matrix T_o^{-1} is built by using the first two rows of M_o, which are linearly independent, and then adding a third row selected such that it is linearly independent from the first two. In this way, we obtain

$$T_o^{-1} = \begin{bmatrix} 0 & 1 & 0 \\ 1 & 1 & 0 \\ -1 & 0 & 1 \end{bmatrix} \quad (9.24)$$

Using Eqs. (5.7), we derive the state matrices of the observability canonical form:

$$\hat{A} = T_o^{-1} A T_o = \begin{bmatrix} 0 & 1 & 0 \\ -2 & 1 & 0 \\ 0.5 & 0.5 & -1 \end{bmatrix}; \quad \hat{B} = T_o^{-1} B = \begin{bmatrix} 3 \\ 5 \\ 2 \end{bmatrix}; \quad (9.25)$$
$$\hat{C} = C T_o = \begin{bmatrix} 1 & 0 & 0 \end{bmatrix}; \quad D = 0$$

The observable part is of order two, while the unobservable part of order one.

A system that is completely observable can be written in *observability canonical form*. This state-space representation, which we write for the SISO case, is defined by a special form for the matrices A and C, as follows:

$$A^{(o)} = \begin{bmatrix} 0 & 0 & \cdots & 0 & -\alpha_n \\ 1 & 0 & \cdots & 0 & -\alpha_{n-1} \\ 0 & 1 & \cdots & 0 & -\alpha_{n-2} \\ \vdots & & & & \vdots \\ 0 & 0 & \cdots & 1 & -\alpha_1 \end{bmatrix}; \quad B^{(o)} = \begin{bmatrix} b_n \\ b_{n-1} \\ b_{n-2} \\ \vdots \\ b_1 \end{bmatrix}; \quad (9.26)$$
$$C^{(o)} = \begin{bmatrix} 0 & 0 & \cdots & 0 & 1 \end{bmatrix}; \quad D^{(o)} = d$$

where, as for the reachability canonical form, $\alpha_1, \alpha_2, \ldots, \alpha_n$ and $b_1, b_2, \ldots, b_n$ are the coefficients of the polynomials in the denominator and numerator of the transfer function:

$$G(s) = \frac{b_1 s^{n-1} + \ldots + s b_{n-1} + b_n}{s^n + \alpha_1 s^{n-1} + \ldots + \alpha_n} + d \quad (9.27)$$

The matrix T appearing in the state transformation $\tilde{x} = T^{-1} x$ is constructed based on the relationship between M_o and $M_o^{(o)}$. Using Eqs. (5.7) in the expression of $M_o^{(o)}$, we obtain that

$$M_o^{(o)} = \begin{bmatrix} C^{(o)} \\ C^{(o)}A^{(o)} \\ \vdots \\ C^{(o)}(A^{(o)})^{n-1} \end{bmatrix}$$

$$= \begin{bmatrix} CT \\ CTT^{-1}AT \\ \vdots \\ CTT^{-1}A^{n-1}T \end{bmatrix} \tag{9.28}$$

$$= \begin{bmatrix} C \\ CA \\ \vdots \\ CA^{n-1} \end{bmatrix} T$$

$$= M_o T$$

and so

$$T = M_o^{-1}M_o^{(o)} \tag{9.29}$$

Example 9.4 _____

Consider again the continuous-time LTI system with state matrices:

$$A = \begin{bmatrix} 1 & -1 & 3 \\ -2 & 0 & 4 \\ 1 & 1 & -3 \end{bmatrix}; \quad B = \begin{bmatrix} 2 \\ -1 \\ 1 \end{bmatrix}; \quad C = \begin{bmatrix} 0 & 1 & 1 \end{bmatrix}; \quad D = 0 \tag{9.30}$$

whose reachability properties have been studied in Example 9.2.
Since the observability matrix

$$M_o = \begin{bmatrix} C \\ CA \\ \vdots \\ CA^{n-1} \end{bmatrix} = \begin{bmatrix} 0 & 1 & 1 \\ -1 & 1 & 1 \\ -2 & 2 & -2 \end{bmatrix} \tag{9.31}$$

has rank equal to three, the system is also completely observable. To derive the observability canonical form, we recall the transfer function computed in Example 9.2:

$$G(s) = \frac{-2s - 12}{s^3 + 2s^2 - 12s + 8} \tag{9.32}$$

Using the coefficients of the numerator and denominator polynomials of the transfer function, we can obtain the state matrices of the observability canonical form:

$$A^{(o)} = \begin{bmatrix} 0 & 0 & -8 \\ 1 & 0 & 12 \\ 0 & 1 & -2 \end{bmatrix}; \quad B^{(o)} = \begin{bmatrix} -12 \\ -2 \\ 0 \end{bmatrix}; \quad C^{(o)} = \begin{bmatrix} 0 & 0 & 1 \end{bmatrix}; \quad D^{(o)} = 0 \tag{9.33}$$

As a direct consequence of the definition of the reachability and observability matrix, it follows that the reachability matrix of a system with state matrices A, B, and C coincides with the observability matrix of a system with state matrices A^T, C^T, and B^T. Analogously, the observability matrix of a system with state matrices A, B, and C coincides with the reachability matrix of a system with state matrices A^T, C^T, and B^T. For this reason, the two systems are said to be *dual*, and the two properties, reachability and observability, are said to be dual properties.

A form that immediately prompts for assessing the reachability and observability properties is the diagonal form studied in Sec. 7.5, Eqs. (7.33). For this state representation, it is immediate to conclude that the observable modes are those for which $c_i \neq 0$, and the reachable modes are those for which $b_i \neq 0$. This is evident also from the transfer function

$$G(s) = \frac{b_1 c_1}{s - \lambda_1} + \frac{b_2 c_2}{s - \lambda_2} + \ldots + \frac{b_n c_n}{s - \lambda_n} \qquad (9.34)$$

that shows that only the reachable and observable modes, i.e., modes for which $b_i \neq 0$ and $c_i \neq 0$, yield non-zero terms in the transfer function, thus effectively contributing to determining the input-output relationship of the system.

9.3 System decomposition and minimal form

In general, an LTI system can be decomposed into four parts through a linear transformation: a reachable and unobservable part (part a), a reachable and observable part (part b), an unreachable and unobservable part (part c), and an unreachable and an observable part (part d). In this state-space realization, known as the Kalman decomposition, the system equations read:

$$\begin{aligned} \dot{\mathbf{x}} &= \hat{A}\hat{\mathbf{x}} + \hat{B}\mathbf{u} \\ \mathbf{y} &= \hat{C}\hat{\mathbf{x}} + \hat{D}\mathbf{u} \end{aligned} \qquad (9.35)$$

with

$$\hat{A} = \begin{bmatrix} \hat{A}_a & \hat{A}_{ab} & \hat{A}_{ac} & \hat{A}_{ad} \\ 0 & \hat{A}_b & 0 & \hat{A}_{bd} \\ 0 & 0 & \hat{A}_c & \hat{A}_{cd} \\ 0 & 0 & 0 & \hat{A}_d \end{bmatrix}; \quad \hat{B} = \begin{bmatrix} \hat{B}_a \\ \hat{B}_b \\ 0 \\ 0 \end{bmatrix}; \qquad (9.36)$$

$$\hat{C} = \begin{bmatrix} 0 & \hat{C}_b & 0 & \hat{C}_d \end{bmatrix}; \qquad \hat{D} = D$$

The block structure of the state matrices reflects the properties of the different parts composing the system. For instance, since the variables of the reachable and unobservable part (part a) can be influenced by the variables of the other parts, the blocks $\hat{A}_{ab}$, $\hat{A}_{ac}$, and $\hat{A}_{ad}$ are generally different than zero. Conversely, since part b is reachable and observable, then it can be influenced by the variables of part d, which is reachable and unobservable, but not by the variables of parts a and c as they are unobservable. Part c can be affected only by the variables of part d, and part d cannot be affected by the variables of any other part. The matrix $\hat{B}$ is such that non-zero blocks may appear only in correspondence of reachable parts. Analogously, the matrix $\hat{C}$ may have non-zero blocks only in correspondence of observable parts. Finally, $\hat{D}$ has no special structure.

Eqs. (9.35) can be explicitly rewritten as follows

$$
\begin{aligned}
\dot{\hat{x}}_a &= \hat{A}_a\hat{x}_a + \hat{A}_{ab}\hat{x}_b + \hat{A}_{ac}\hat{x}_c + \hat{A}_{ad}\hat{x}_d + \hat{B}_a\mathbf{u} \\
\dot{\hat{x}}_b &= \hat{A}_b\hat{x}_b + \hat{A}_{bd}\hat{x}_d + \hat{B}_b\mathbf{u} \\
\dot{\hat{x}}_c &= \hat{A}_c\hat{x}_c + \hat{A}_{cd}\hat{x}_d \\
\dot{\hat{x}}_d &= \hat{A}_d\hat{x}_d \\
\mathbf{y} &= \hat{C}_b\hat{x}_b + \hat{C}_d\hat{x}_d + \hat{D}\mathbf{u}
\end{aligned}
\tag{9.37}
$$

In this lecture, we do not delve into the construction of the state transformation matrix T used for the Kalman decomposition. Instead, we refer to the use of the MATLAB command `minreal` for deriving the decomposition, as illustrated in Sec. 9.4.

A fundamental property of the transfer function derives from Eqs. (9.35) and (9.36). The transfer function (or the transfer matrix in the MIMO case) depends exclusively on the reachable and observable part of the system. This is because the transfer function represents the input-output relationship of a system, which is determined solely by the components affected by the input and that influence the output. By direct calculation, in fact, one finds that

$$
G(s) = \hat{C}_b\left(s\mathrm{I} - \hat{A}_b\right)^{-1}\hat{B}_b + \hat{D}
\tag{9.38}
$$

A system that is completely reachable and observable is said to be in minimal form. Conversely, the reachable and observable part of a system is referred to as the minimal realization, as it represents the smallest subsystem required to capture the system's input-output relationship.

Example 9.5 _____

Consider the continuous-time LTI system with state matrices

$$
\hat{A} = \begin{bmatrix} 3 & 0 & 0 & 0 \\ 0 & -2 & 1 & 0 \\ 0 & -1 & -2 & 0 \\ 0 & 0 & 0 & -5 \end{bmatrix} ; \quad \hat{B} = \begin{bmatrix} 1 \\ 0 \\ 0 \\ 0 \end{bmatrix} ;
\tag{9.39}
$$

$$
\hat{C} = \begin{bmatrix} 1 & 0 & 0 & 1 \end{bmatrix} ; \quad \hat{D} = 0
$$

This is in Kalman form, where the first part is of order 1 and is reachable and observable, the second part is of order 2 and is unreachable and unobservable, and the third part is unreachable and observable. The transfer function depends only on the reachable and observable part and is equal to $G(s) = \frac{1}{s-3}$.

9.4 Reachability and observability in MATLAB

The reachability and observability matrices can be calculated in MATLAB®, using the commands `ctrb(A,B)` and `obsv(A,B)`. We now illustrate their use with reference to the examples discussed in the lecture.

MATLAB® exercise 9.1 _____

We begin with Example 9.1 and show the MATLAB® commands to isolate the reachable and non-reachable parts of the system.
First, the state matrices of the system are defined:

```
A=[0  1  0; -1  1  0; 2 -1 -1];
B=[0  1 -1]';
C=[0  0  1];
D=0;
```

Next, we compute the reachability matrix, its determinant and its rank:

```
Mr=ctrb(A,B)
det(Mr)
rank(Mr)
```

We proceed by constructing the matrix T_r as follows:

```
Tr=zeros(3);
Tr(:,1:2)=Mr(:,1:2)
Tr(:,3)=[0  0  1]'
```

Lastly, we compute the state-space representation (9.5):

```
Ahat=inv(Tr)*A*Tr
Bhat=inv(Tr)*B
Chat=C*Tr
Dhat=D
```

MATLAB® exercise 9.2 _____

Consider now the problem in Example 9.2 dealing with the reachability canonical form of a system. First, we define the state matrices of the system:

```
A=[1 -1  3; -2  0  4; 1  1 -3];
B=[2 -1  1]';
C=[0  1  1];
D=0;
```

Next, we calculate the reachability matrix and its determinant:

```
Mr=ctrb(A,B)
det(Mr)
```

Since it is different than zero, the system is completely reachable. To derive the reachability canonical form, we first compute the transfer function

```
system1=ss(A,B,C,D)
tf(system1)
```

and obtain $G(s) = \frac{-2s-12}{s^3+2s^2-12s+8}$. From this expression, we write the state matrices of the reachability canonical form:

```
AR=diag([1 1],1);
AR(3,:)=[-8 12 -2]
BR=[0; 0; 1]
CR=[-12 -2 0]
DR=0;
```

To verify, we can compute the transfer function and confirm that it matches the one we started with

```
system1R=ss(AR,BR,CR,DR)
tf(system1R)
```

Next, we perform an additional verification by computing the matrix T that represents the state transformation from one representation to another:

```
MrR=ctrb(AR,BR);
T=Mr*inv(MrR)
```

We then verify that the state-space matrices obtained using Eqs. (5.7) correspond to those of the reachability canonical form:

```
AR=inv(T)*A*T
BR=inv(T)*B
CR=C*T
DR=D
```

However, note that these steps serve as a verification of the procedure, but cannot be used for deriving the reachability canonical form, as computing T requires the reachability matrix $M_r^{(r)}$ of the reachability canonical form.

MATLAB® exercise 9.3

Here, we illustrate the commands related to Example 9.3. First, the state matrices of the system are defined:

```
A=[0 -1 1; 1 0 1; 1 -1 0];
B=[1 -1 2]';
C=[1 0 1];
D=0;
```

Next, we compute the observability matrix, its determinant and rank

```
Mo=obsv(A,C)
det(Mo)
rank(Mo)
```

We proceed by constructing the matrix T_o^{-1} as follow:

```
ToInv=zeros(3);
ToInv(1:2,:)=Mo(1:2,:)
ToInv(3,:)=[0 0 1]
```

Lastly, we compute the state-space representation (9.21):

```
To=inv(ToInv)
Ahat=inv(To)*A*To
Bhat=inv(To)*B
Chat=C*To
Dhat=D
```

MATLAB® exercise 9.4 _____

Consider now Example 9.4, which deals with the observability canonical form of a system.
We start by defining the state matrices of the system:

```
A=[1 -1 3; -2 0 4; 1 1 -3];
B=[2 -1 1]';
C=[0 1 1];
D=0;
```

Next, we calculate the observability matrix and its determinant

```
Mo=obsv(A,C)
det(Mo)
```

The determinant is different than zero. This yields that the system is completely observable and we can write the observability canonical form. To do this, we start with the transfer function:

```
system1=ss(A,B,C,D)
tf(system1)
```

that is $G(s) = \frac{-2s-12}{s^3+2s^2-12s+8}$. From this expression, we write the state matrices of the observability canonical form:

```
Ao=diag([1 1],-1);
Ao(:,3)=[-8 12 -2]
Bo=[-12; -2; 0]
Co=[0 0 1]
Do=0;
```

and check that the transfer function matches the one we derived earlier:

```
system1o=ss(Ao,Bo,Co,Do)
tf(system1o)
```

Lastly, we compute the matrix T as follows:

```
Moo=obsv(Ao,Co);
T=inv(Mo)*Moo
```

and verify that the state-space matrices obtained using Eqs. (5.7) correspond to those of the observability canonical form:

```
Ao=inv(T)*A*T
Bo=inv(T)*B
Co=C*T
Do=D
```

Also in this case, we remark that these steps serve as a verification of the procedure, but not for deriving the observability canonical form, as computing T requires the observability matrix $M_o^{(o)}$.

Lastly, we show how to obtain the Kalman decomposition in MATLAB®. To do this, the command **minreal** can be used. This command allows one to calculate the minimal realization of a system. At the same time, it returns the transformation matrix to derive the Kalman decomposition. Specifically, assumed that we want to calculate the Kalman composition of the system called **system1**, we can use the following command:

```
[sys1mr, U]=minreal(system1)
```

Here, `sys1mr` is the minimal realization, and `U` the inverse of the transformation matrix T. The next example illustrates the use of this command.

MATLAB® exercise 9.5 _____

Let us first define the system, by giving its state matrices

```
A=[3.5 2 -1 -6; -4.875 -7 14.75 -5; 3.75  0 2.5 -8; 2.625 1 -0.25 -5];
B=[0.5 1.875 1.25  0.375]';
C=[2 0 2 -4];
D=0;
system1=ss(A,B,C,D);
```

As a first step, we calculate the transfer function using the command `zpk` that factorize the numerator and denominator polynomials:

```
zpk(system1)
```

We obtain $G(s) = \frac{2(s+4)(s+2)^2}{(s-2)(s+2)^2(s+4)}$, which shows that there are certain cancellations between some zeros and some poles of the system. After the cancellations, we find that the reachable and observable part is of order 1. Now, we apply the `minreal` command and, then, we find the state matrices in the Kalman form:

```
[sys1mr , U]=minreal(system1)
T=inv(U);
Ahat=inv(T)*A*T
Bhat=inv(T)*B
Chat=C*T
Dhat=D
```

We obtain the following state matrices:

$$
\hat{A} = \begin{bmatrix} 2.0000 & 0.0000 & -0.0000 & 1.6036 \\ 14.3183 & -2.0000 & -0.0000 & -10.6778 \\ 6.6852 & 0.5011 & -2.0000 & 9.5551 \\ 0.0000 & -0.0000 & 0.0000 & -4.0000 \end{bmatrix}; \quad \hat{B} = \begin{bmatrix} -0.4226 \\ -2.1958 \\ -0.6847 \\ 0.0000 \end{bmatrix}; \quad (9.40)
$$

$$
\hat{C} = \begin{bmatrix} -4.7329 & -0.0000 & 0.0000 & -1.2649 \end{bmatrix}; \quad \hat{D} = 0
$$

This system is in Kalman form with a different ordering of the parts with respect to Eqs. (9.35) and (9.36). Here the first part is of order 1 and is reachable and observable. The second part is of order 2 and is reachable and unobservable, and the third part is of order 1 and is unreachable and unobservable. There is no unreachable and observable part in this system.

9.5   Exercises

1. Find γ such that the continuous-time LTI system with state matrices

$$
A = \begin{bmatrix} 0 & -1 \\ 4 & \gamma \end{bmatrix}; \quad B = \begin{bmatrix} -2 \\ 1 \end{bmatrix}; \quad (9.41)
$$

$$
C = \begin{bmatrix} 3 & -5 \end{bmatrix}; \quad D = 2
$$

is completely controllable.

2. Determine, if possible, the reachability canonical form of the continuous-time LTI system with the following state matrices

$$A = \begin{bmatrix} 1 & -1 & 2 \\ 0 & 1 & 0 \\ -2 & 3 & 10 \end{bmatrix}; \quad B = \begin{bmatrix} -1 \\ -1 \\ 1 \end{bmatrix};$$
$$C = \begin{bmatrix} 1 & 0 & 0 \end{bmatrix}; \quad D = 0$$
(9.42)

3. Calculate the observability matrix for the continuous-time LTI system with state matrices

$$A = \begin{bmatrix} -1 & -1 \\ 1 & -1 \end{bmatrix}; \quad B = \begin{bmatrix} 0 \\ 1 \end{bmatrix};$$
$$C = \begin{bmatrix} 1 & 1 \end{bmatrix}; \quad D = 0$$
(9.43)

and study the observability property of the system.

4. Determine, if possible, the observability canonical form of the continuous-time LTI system with the following state matrices

$$A = \begin{bmatrix} 2 & -1 & -5 \\ 1 & 0 & 1 \\ -1 & -1 & 0 \end{bmatrix}; \quad B = \begin{bmatrix} 1 \\ 0 \\ 2 \end{bmatrix};$$
$$C = \begin{bmatrix} 0 & 1 & 1 \end{bmatrix}; \quad D = 0$$
(9.44)

5. Using MATLAB®, calculate the Kalman decomposition and the transfer function of the system with state matrices:

$$A = \begin{bmatrix} -3 & 8 & -\frac{4}{3} & -7 \\ 1 & -3 & \frac{23}{3} & 0 \\ 0 & 3 & 5 & -3 \\ 1 & 2 & \frac{23}{3} & -5 \end{bmatrix}; \quad B = \begin{bmatrix} \frac{2}{9} \\ \frac{5}{9} \\ \frac{1}{3} \\ \frac{5}{9} \end{bmatrix};$$
$$C = \begin{bmatrix} 0 & 3 & 3 & -3 \end{bmatrix}; \quad D = 0$$
(9.45)

10

Lecture 10—Time-domain specifications and behavior of first-order and second-order systems

---

In this lecture, we study the step response of first-order and second-order continuous-time LTI systems. We define several indicators for a quantitative comparison between the output of different systems and show how these are linked to the parameters of the system poles. Next, we introduce the notion of dominant poles and show how they can be used to approximate a system by a lower-order model. In this way, higher-order systems can be approximately represented as first-order systems, second-order systems, or, eventually, as a combination of the two.

10.1 Time-domain specifications

In order to compare the behavior of different systems, we need quantitative parameters that provide information on how the system responds to inputs. This is particularly relevant when a controller is applied to the system, as these parameters allow us to compare the effects of different control laws (or control parameters) on the system output. To this aim, it is convenient to refer to the step response of a BIBO stable system. More specifically, we will study first-order and second-order systems, and we will show that the results we derive are important also to characterize higher-order systems.

Let us consider the step response shown in Fig. 10.1 and let us indicate with y_∞ the steady-state value of $y(t)$, namely $y_\infty = \lim_{t \to +\infty} y(t)$, and with $y_{\max}$ the maximum value taken by $y(t)$, namely $y_{\max} = \max_t y(t)$. Several parameters can be used to quantify the characteristics of this response:

- the *rise time* t_r. It quantifies how quickly the system responds to a solicitation. It is defined as the time needed by the system to move from 10% of the steady-state value to 90% of it, namely to move from $0.1y_\infty$ to $0.9y_\infty$;

- the *settling time* $t_{s,\varepsilon}$. It quantifies how quickly the system converges to the steady-state value. Given $\varepsilon > 0$, it is defined as the minimum time after

DOI: 10.1201/9781003487289-10

which the output remains confined in a range of $\pm\varepsilon\%$ of the steady-state value y_∞, namely as the minimum time $t_{s,\varepsilon}$ such that, for any $t > t_{s,\varepsilon}$, then $(1 - 0.01\epsilon) < y(t) < (1 + 0.01\epsilon)$;

- the *percentage overshoot* $s_\%$. It quantifies in percentage how much the maximum peak of $y(t)$ surpasses the steady-state value y_∞. It is defined as $s_\% = \frac{y_{\max} - y_\infty}{y_\infty} \cdot 100$.

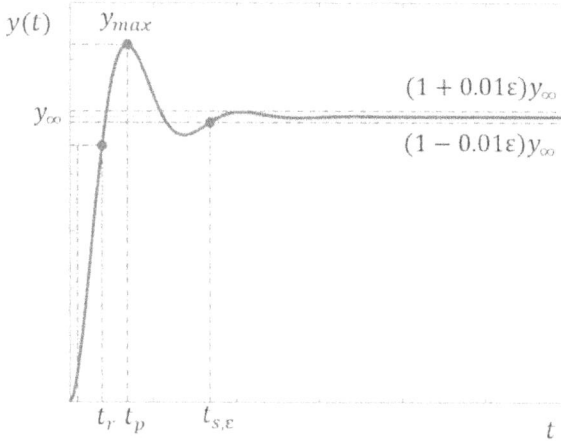

FIGURE 10.1
Step response of a BIBO stable system.

In summary, the rise time provides a measure of the speed of the system response, the settling time quantifies the duration of the transient response, and the overshoot indicates the maximum deviation during the transient. Often, reducing the overshoot increases the rise time, so to avoid an excessively slow response, a certain level of overshoot is often acceptable.

10.2   First-order systems

We consider first-order systems with the following transfer function:

$$G(s) = \frac{\mu}{1 + Ts} \tag{10.1}$$

where $T > 0$ is the time constant associated with the system pole $s = -1/T$, and $\mu > 0$ is the static gain.

Example 10.1 _____

The model of cruise speed control studied in Example 2.1 provides an example of a first-order system. In fact, its governing equation

$$\dot{v} + \frac{b}{m}v = \frac{u}{m} \tag{10.2}$$

yields the following transfer function

$$G(s) = \frac{V(s)}{U(s)} = \frac{b}{1 + s\frac{m}{b}} \tag{10.3}$$

that corresponds to Eq. (10.1) with $\mu = b$ and $T = \frac{m}{b}$.

Example 10.2 _____

Another example of first-order system is the electrical circuit studied in Example 2.6. In fact, the circuit is governed by the following equation

$$\dot{x} = -\frac{1}{RC}x + \frac{1}{RC}u \tag{10.4}$$

while the transfer function is

$$G(s) = \frac{Y(s)}{U(s)} = \frac{1}{1 + sRC} \tag{10.5}$$

that corresponds to Eq. (10.1) with $\mu = 1$ and $T = RC$.

For first-order systems described by the transfer function (10.1), the unit step response is given by

$$y(t) = \mu \left(1 - e^{-\frac{t}{T}}\right) \tag{10.6}$$

The output $y(t)$ is exemplified in Fig. 10.2 for a system with $\mu = 1$ and $T = 5$. It has a monotonic trend, such that there are no oscillations and no overshoot. The settling time can be calculated from the condition:

$$e^{-\frac{t_{s,\varepsilon}}{T}} = \frac{\varepsilon}{100} \tag{10.7}$$

which gives

$$t_{s,\varepsilon} = -T\ln(0.01\varepsilon) \tag{10.8}$$

Specifically, when $\varepsilon = 1$, we have

$$t_{s,1\%} \simeq 4.6T \tag{10.9}$$

For the rise time, with similar arguments it can be derived that

$$t_r \simeq 2.2T \tag{10.10}$$

The rise time t_r and the settling time $t_{s,\varepsilon}$ depend exclusively on the time constant T. The larger is T, the larger are t_r and $t_{s,\varepsilon}$.

Step Response

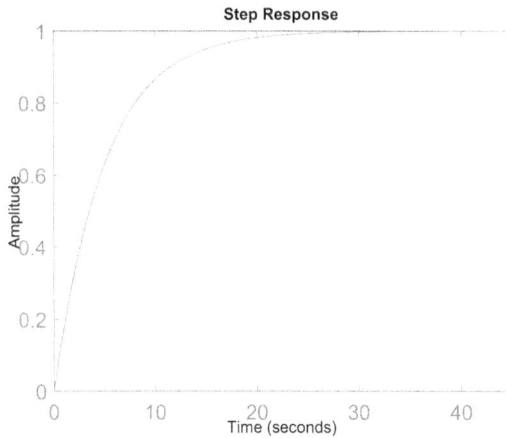

FIGURE 10.2
Step response of a first-order system $G(s) = \frac{\mu}{1+Ts}$ with $\mu = 1$ and $T = 5$.

10.3   Second-order systems

For second-order systems we assume the following generic transfer function

$$G(s) = \frac{\mu\omega_n^2}{s^2 + 2\xi\omega_n s + \omega_n^2} \tag{10.11}$$

where, also in this case, we restrict the analysis to BIBO stable systems, that is, $\xi > 0$. Notice that the system poles can be either real, if $\xi \geq 1$, or complex and conjugate, if $0 < \xi < 1$.

Example 10.3 _____

The mechanical system Fig. 10.3 is an example of a second-order system. The Newton's law for this system yields:

$$m\ddot{x} + b\dot{x} + kx = u \tag{10.12}$$

Applying the Laplace transform with zero initial conditions we obtain the transfer function of the system:

$$G(s) = \frac{X(s)}{U(s)} = \frac{\frac{1}{m}}{s^2 + \frac{b}{m}s + \frac{k}{m}} \tag{10.13}$$

This transfer function corresponds to Eq. (10.11) with $\mu = \frac{1}{k}$, $\omega_n = \sqrt{\frac{k}{m}}$, and $\xi = \frac{b}{2\sqrt{mk}}$. The poles are real if $\xi \geq 1$, that is, if $b^2 - 4km \geq 0$. Instead, they are complex and conjugate if $b^2 - 4km < 0$.

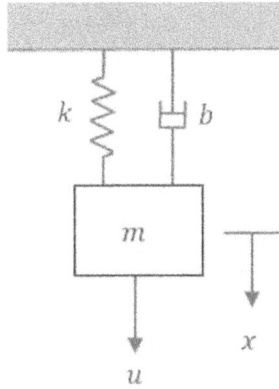

FIGURE 10.3
An example of a second-order linear system.

Example 10.4 _____

Another example of a second-order system is the electrical circuit in Examples 2.7 and 7.4, where we have found that the transfer function is

$$G(s) = \frac{\frac{1}{LC}}{s^2 + \frac{R}{L}s + \frac{1}{LC}} \tag{10.14}$$

This transfer function corresponds to Eq. (10.11) with $\mu = 1$, $\omega_n = \frac{1}{\sqrt{LC}}$, and $\xi = \frac{R}{2}\sqrt{\frac{C}{L}}$.

The poles are real if $\xi \geq 1$, that is, if $R \geq 2\sqrt{\frac{L}{C}}$. Instead, they are complex and conjugate if $R < 2\sqrt{\frac{L}{C}}$.

Let us now consider the case where the system (10.11) has real poles. In this scenario, the transfer function can be expressed in the following form:

$$G(s) = \frac{\mu}{(1 + T_1 s)(1 + T_2 s)} \tag{10.15}$$

with $T_1 > T_2 > 0$. The unit step response is given by

$$y(t) = \mu \left(1 - \frac{T_1}{T_1 - T_2} e^{-\frac{t}{T_1}} + \frac{T_2}{T_1 - T_2} e^{-\frac{t}{T_2}} \right) \tag{10.16}$$

This expression does not prompt for a simple relationship between t_r, $t_{s,\varepsilon}$, $s_\%$, and T_1 and T_2. On the other hand, the response is monotonic and not particularly fast, making this condition often undesirable to achieve with control (e.g., Fig. 10.4, which refers to a system with $T_1 = 5$ and $T_2 = 10$).

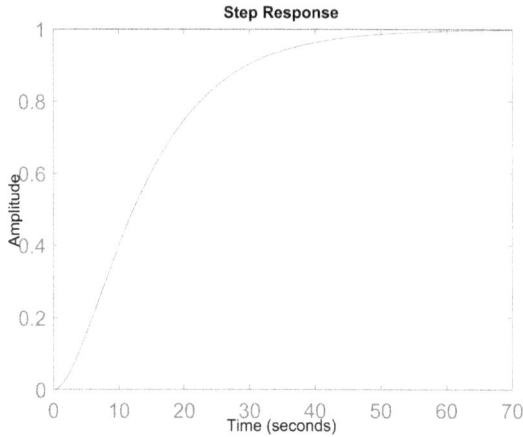

FIGURE 10.4
Step response of $G(s) = \frac{\mu}{(1+T_1 s)(1+T_2 s)}$ with $T_1 = 5$ and $T_2 = 10$.

Notice that, if $T_1 \gg T_2$, then $y(t) \simeq \mu \left(1 - e^{-\frac{t}{T_1}}\right)$, and the system behaves as a first-order one. Therefore, the relationships found for a first-order system, namely (10.9) and (10.10), can be used.

In the particular case that $T_1 = T_2$, the two poles are coincident. The transfer function reads

$$G(s) = \frac{\mu}{(1 + Ts)^2} \tag{10.17}$$

and the unit step response is given by

$$y(t) = \mu \left(1 - e^{-\frac{t}{T_1}} + \frac{t}{T} e^{-\frac{t}{T}}\right) \tag{10.18}$$

Also, in this case, the step response is monotone. Suitable approximated values for $t_{s,1\%}$ and t_r are

$$t_{s,1\%} \simeq 6.64T \tag{10.19}$$

and

$$t_r \simeq 3.36T \tag{10.20}$$

Notice that these times are larger than the corresponding values for first-order systems, as in (10.9) and (10.10).

The second case we analyze is a system with complex and conjugate poles, described by the following transfer function:

$$G(s) = \frac{\mu \omega_n^2}{s^2 + 2\xi \omega_n s + \omega_n^2} \tag{10.21}$$

with $0 < \xi < 1$ and $\omega_n > 0$. The system poles are $s_{1,2} = -\xi \omega_n \pm j\omega_n \sqrt{1 - \xi^2}$.

Fig. 10.5 illustrates the position in the complex plane of one of the two conjugate poles (the one with a positive imaginary part). Here, we notice that the modulus of the pole is ω_n (in fact $\sqrt{(\xi \omega_n)^2 + (\omega_n \sqrt{1 - \xi^2})^2} = \omega_n$) and the angle θ is equal to $\theta = \arcsin \xi$.

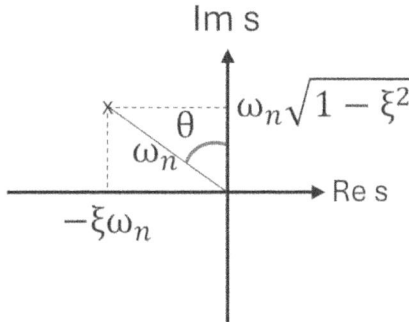

FIGURE 10.5
Position in the complex plane of one of the two conjugate poles of $G(s) = \frac{\mu \omega_n^2}{s^2 + 2\xi \omega_n s + \omega_n^2}$.

The unit step response of $G(s) = \frac{\mu \omega_n^2}{s^2 + 2\xi \omega_n s + \omega_n^2}$ is given by

$$y(t) = \mu \left(1 - \frac{1}{\sqrt{1 - \xi^2}} e^{-\xi \omega_n t} \sin \left(\omega_n \sqrt{1 - \xi^2} t + \arccos \xi \right) \right) \tag{10.22}$$

Fig. 10.6 illustrates the curve $y(t)$ for several values of ξ, also including the case $\xi = 0$. It shows that for $\xi = 0$ oscillations become persistent, while for larger values of ξ the overshoot reduces and the oscillations damp out in shorter times.

An approximated expression for the rise time can be found by considering the average curve $y(t)$ for $\xi = 0.5$:

$$t_r \simeq \frac{1.8}{\omega_n} \tag{10.23}$$

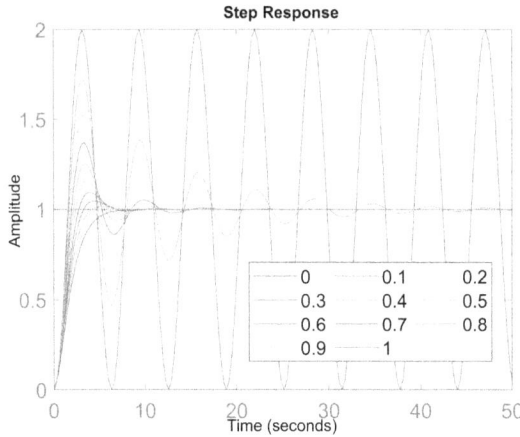

FIGURE 10.6
Step response of a system with transfer function $G(s) = \frac{\mu\omega_n^2}{s^2+2\xi\omega_n s+\omega_n^2}$ for different values of ξ. Here $\mu = 1$ and $\omega_n = 1$.

The time of the first peak is given by

$$t_p = \frac{\pi}{\omega_n\sqrt{1-\xi^2}} \tag{10.24}$$

that, replaced in the expression of $y(t)$, allows one to calculate the overshoot:

$$s_\% = 100e^{-\frac{\pi\xi}{\sqrt{1-\xi^2}}} \tag{10.25}$$

The expression $s_\% = 100e^{-\frac{\pi\xi}{\sqrt{1-\xi^2}}}$ is plotted against ξ in Fig. 10.7 that can be used in place of the formula to obtain approximated values of ξ guaranteeing a given $s_\%$. For instance, $s_\% < 10\%$ can be obtained by selecting $\xi \geq 0.6$.

Finally, the condition $e^{-\xi\omega_n t_s} = 0.01\varepsilon$ allows to calculate the settling time. For instance for $t_{s,1\%}$ we have

$$t_{s,1\%} \simeq \frac{4.6}{\xi\omega_n} \tag{10.26}$$

10.4   Controlling the characteristics of the system output

The relationships discussed in the previous sections are useful to control the characteristics of the output of a system. The most general case is represented by the second-order system with complex and conjugate poles, as the response

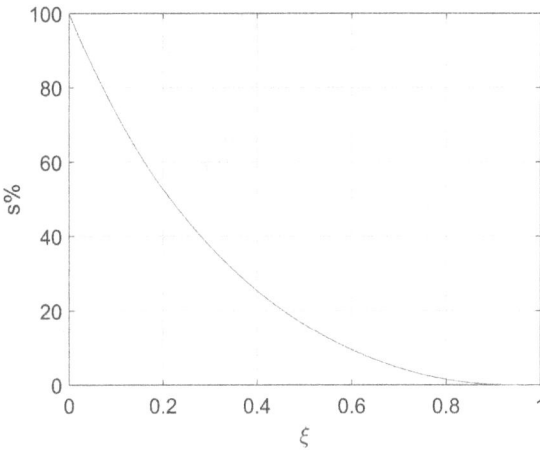

FIGURE 10.7
Percentage overshoot $s_\%$ vs. ξ, as per Eq. (10.25).

can display overshoot and damped oscillations. Therefore, we can refer to the relationships (10.23), (10.25), and (10.26) to determine which ω_n and ξ can guarantee given performance in the system output.

Suppose, for instance, that a settling time smaller than a given value is required, then, from (10.26), the product $\xi\omega_n$ needs to be selected as:

$$\xi\omega_n > \frac{4.6}{t_{s,1\%}} \tag{10.27}$$

where $t_{s,1\%}$ is the maximum settling time allowed for the system to control. The constraint (10.27) is visually represented in Fig. 10.8(b). It defines the shadowed region where the poles of the controlled system can be located in order to guarantee that its settling time will be smaller than the assigned value.

Similarly, if a rise time smaller than a given value is required, then, from (10.23), ω_n needs to be selected as:

$$\omega_n > \frac{1.8}{t_r} \tag{10.28}$$

This constraint is visually represented in Fig. 10.8(b).

For the overshoot, Eq. (10.25) and Fig. 10.7 express its relationship with ξ. Let $\xi(s_\%)$ indicate the inverse relationship between the two variables, then in order to have an overshoot smaller than a given value $\bar{s}_\%$ one has to consider

$$\xi > \xi(\bar{s}_\%) \tag{10.29}$$

Taking into account that $\xi = \sin\theta$ (see Fig. 10.5), we have that the region in the complex plane where $s_\% < \bar{s}_\%$ is defined by the condition

$$\theta > \arcsin(\xi(\bar{s}_\%)) \tag{10.30}$$

This region is schematically illustrated in Fig. 10.8(c).

Considering altogether the requirements on the three parameters $t_{s,1\%}$, t_r, and $s_\%$, we obtain the region illustrated in Fig. 10.8(d), which represent an area where the poles of the system can be located to guarantee the desired performance. However, it is important to remark that this is rigorously true only if the system is second-order and has no zeros. Otherwise, the relationships we have illustrated should be considered as general guidelines, rather than exact conditions. As an example, in Secs. 10.5 and 10.6 we will see how, in the presence of a zero or of an additional pole in the transfer function, the response deviates from the original one.

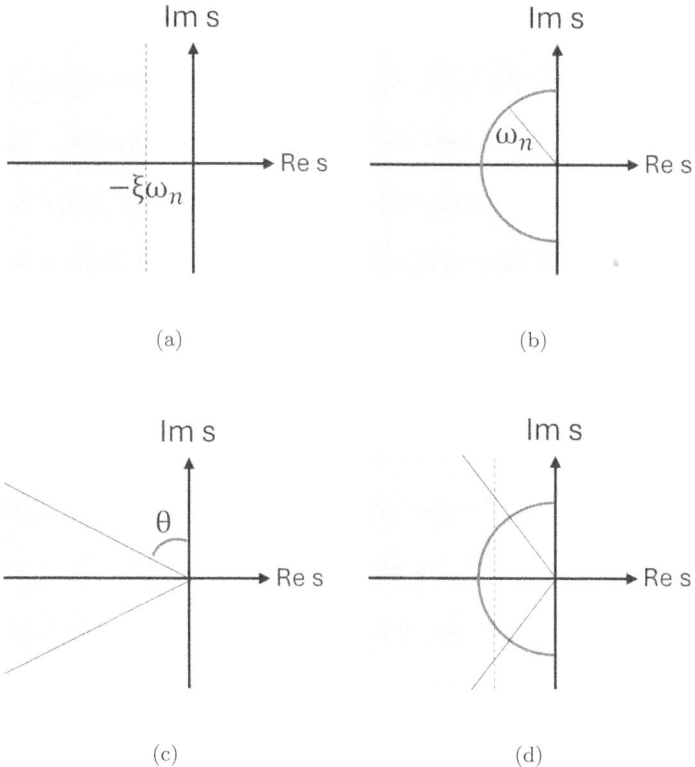

FIGURE 10.8
Regions where the system poles should be located in order to guarantee desired performance on: (a) $t_{s,1\%}$; (b) t_r; (c) $s_\%$; (d) all three parameters.

Example 10.5 _____

Suppose we want to design a system with $s_\% \leq 10\%$, $t_r \leq 2$ s, and $t_{s,1\%} \leq 4$ s. From (10.23) we derive that $\omega_n \geq 1.8/2 = 0.9$ rad/s. Moreover, from (10.26) we get that $\xi\omega_n > 4.6/4 = 1.15$ rad/s; fixing $\xi = 0.6$, we obtain $\omega_n \geq 4.6/2.4 = 1.92$ rad/s. To meet these constraints, we can select, for instance, $\xi = 0.6$ and $\omega_n = 2$ rad/s.

The analysis of the second-order systems studied in this section indicates that decreasing the rise time requires increasing ω_n, while reducing the overshoot requires increasing ξ. Though qualitative, these insights can provide important guidelines for control design.

10.5 Effect of a zero

In this section, we study the effect of a zero on the step response of a second-order systems with complex and conjugates poles. To this aim, we analyze the following model:

$$G(s) = \frac{\mu\left(\frac{s}{\alpha\xi\omega_n} + 1\right)}{\frac{s^2}{\omega_n^2} + \frac{2\xi s}{\omega_n} + 1} \qquad (10.31)$$

where α parametrizes the relative distance of the zero from the real part of the poles (which is given by $-\xi\omega_n$).

First of all, we notice that for $\alpha \gg 1$, then $G(s) \simeq \frac{\mu\omega_n^2}{s^2 + 2\xi\omega_n s + \omega_n^2}$. We conclude that a zero located far from the poles has no effect on the system response.

Next, we consider smaller values of α and write $G(s)$ in Eq. (10.31) as the sum of two terms:

$$G(s) = \frac{\mu}{\frac{s^2}{\omega_n^2} + \frac{2\xi s}{\omega_n} + 1} + \frac{1}{\alpha\xi\omega_n}\frac{\mu s}{\frac{s^2}{\omega_n^2} + \frac{2\xi s}{\omega_n} + 1} \qquad (10.32)$$

that shows how the step response is given by two terms: one, associated with the first term in (10.31), which is exactly the step response in the absence of any zero, and one, associated with the first term in (10.31), which is the derivative of the step response in the absence of any zero, weighted by the factor $1/\alpha\xi\omega_n$. In fact, the second term differs from the first by the weighting factor $1/\alpha\xi\omega_n$ and by the presence of s (recall that $sF(s)$ is the Laplace transform of $df(t)/dt$).

Consequently, for $\alpha > 0$ the step response shows a larger overshoot and a smaller rise time. This is illustrated in Fig. 10.9, where the step response for two different values of α ($\alpha = 1$ and $\alpha = 2$) is shown, along with the step response in the case where $G(s)$ has no zero ($\alpha = \infty$) and its derivative. For the sake of illustration, the other parameters have been selected as follows:

$\mu = 1$, $\omega_n = 1$, and $\xi = 0.3$. We notice that the smaller is α, the larger is the effect of the zero on the step response, e.g., it produces a larger overshoot.

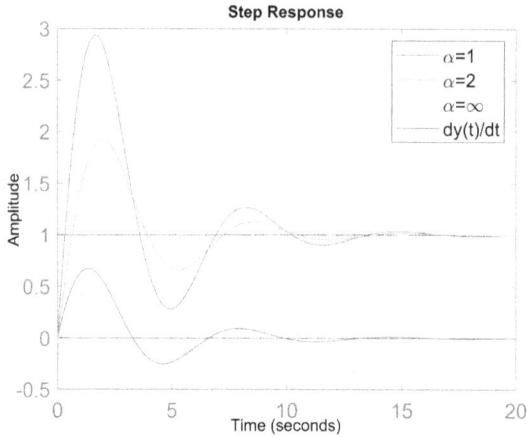

FIGURE 10.9
Effect of a negative zero on the step response of a second-order system.

The case with $\alpha < 0$ is also interesting. Fig. 10.10 illustrates the step response for two values of α ($\alpha = -1$ and $\alpha = -2$) with all other parameters set as in the previous example, namely $\mu = 1$, $\omega_n = 1$, and $\xi = 0.3$. The relevant feature is the presence of a *negative overshoot*. The step response starts with negative values and then moves toward positive values. This results in a longer transient before reaching the steady-state. In addition, there are cases when the negative overshoot can damage the control equipments.

10.6   Effect of an additional pole

To study the effect of an additional pole, we refer to the following transfer function:

$$G(s) = \frac{\mu}{\left(\frac{s}{\alpha\xi\omega_n} + 1\right)\left(\frac{s^2}{\omega_n^2} + \frac{2\xi s}{\omega_n} + 1\right)} \qquad (10.33)$$

The effect of the additional pole is illustrated in Fig. 10.11 where two values of α ($\alpha = 1$ and $\alpha = 2$) are considered (the other parameters are set as follows: $\mu = 1$, $\omega_n = 1$, and $\xi = 0.3$). Also in this case, for very large values of α the effect is negligible, as the additional pole is far from the two other poles of the system. For smaller values, the main effect of the pole is slowing

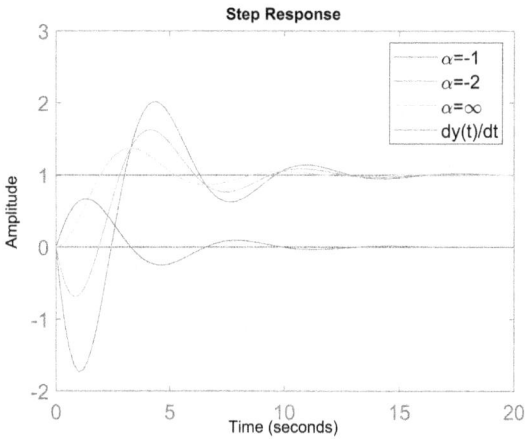

FIGURE 10.10
Effect of a positive zero on the step response of a second-order system.

down the system dynamics, thus yielding a larger rise time. When the pole is close to the other two poles of the system, the shape of the step response is also significantly affected by its presence.

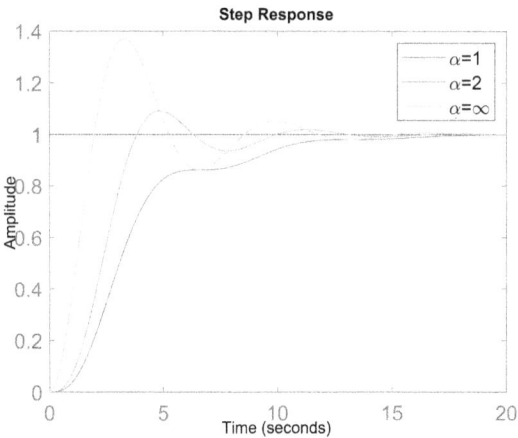

FIGURE 10.11
Effect of an additional pole on the step response of a second-order system.

10.7 Dominant poles

All the analysis previously discussed is important not only for first-order and second-order systems but also for higher-order systems. In fact, there are situations where the behavior of higher-order systems closely resembles that of a lower-order systems, e.g., a first-order or a second-order one. This is the case when one or more poles are *dominant*.

Assuming that the system is BIBO stable, a pole or a pair of poles is said dominant when it is significantly closer to the origin than the other poles, and there are no zeros close to it. In this case, we can assume that the slowest part of the system dominates the response, while the fast part can be neglected. Using this approximation makes possible to approximate a second-order system by a first-order one, and a system of order $n \geq 3$ by either a first-order or a second-order one.

Consider, for instance, the second-order system $G(s) = \frac{\mu p_1 p_2}{(s+p_1)(s+p_2)}$ with $p_1, p_2 > 0$, then if $p_1 \ll p_2$, the pole $s = -p_1$ is much closer to the origin than $s = -p_2$ (in practice, $p_2/p_1 > 5 \div 10$ suffices for this assumption), and $G(s) \simeq \frac{\mu p_1}{s+p_1}$. Vice versa, if $p_2 \ll p_1$, the pole $s = -p_2$ is much closer to the origin than $s = -p_1$, and $G(s) \simeq \frac{\mu p_2}{s+p_2}$. In all other cases, i.e., when p_1 and p_2 have similar order of magnitude, we cannot approximate $G(s)$ with a dominant pole approximation. In summary, we have that

$$G(s) = \frac{\mu p_1 p_2}{(s+p_1)(s+p_2)} \simeq \begin{cases} \frac{\mu p_1}{s+p_1}, & p_1 \ll p_2 \\ \frac{\mu p_2}{s+p_2}, & p_2 \ll p_1 \end{cases} \tag{10.34}$$

Consider now the third-order system $G(s) = \frac{\mu p \omega_n^2}{(s+p)(s^2+2\xi\omega_n s \omega_n^2)}$, in this case, depending on the location of its poles, we may approximate it either as a first-order or a second-order system:

$$G(s) = \frac{\mu p \omega_n^2}{(s+p)(s^2+2\xi\omega_n s \omega_n^2)} \simeq \begin{cases} \frac{\mu p}{s+p}, & p \ll \xi\omega_n \\ \frac{\mu \omega_n^2}{s^2+2\xi\omega_n s \omega_n^2}, & \xi\omega_n \ll p \end{cases} \tag{10.35}$$

Example 10.6 _____

Consider the system with the following transfer function

$$G(s) = \frac{225p}{(s+p)(s^2+0.6s+1)(s^2+6s+225)} \tag{10.36}$$

where $p > 0$. The system has a pair of complex and conjugate poles with $\xi = 0.3$ and $\omega_n = 1$, another pair of complex and conjugates poles with $\xi = 0.2$ and $\omega_n = 15$, and a real pole $s = -p$. Depending on the value of p, there are three possible scenarios, exemplified in Fig. 10.6. If the pole $s = -p$ is far from the pair of complex and conjugate poles with $\xi = 0.3$ and $\omega_n = 1$, then the system is dominated by this pair of poles. Accordingly, its step response displays the typical features of the step response a second-order system, e.g., Fig. 10.12(a) with $p = -4$. If the pole $s = -p$ is close to the pair of complex and conjugate poles with $\xi = 0.3$ and $\omega_n = 1$, then there are no dominant poles, e.g., the step response shown in Fig. 10.12(b) with $p = -0.4$. Lastly, if the pole $s = -p$ is the closest to the origin, then the system is dominated by this single pole and its step response displays the typical monotonic trend of first-order systems, e.g., Fig. 10.12(c) with $p = -0.08$.

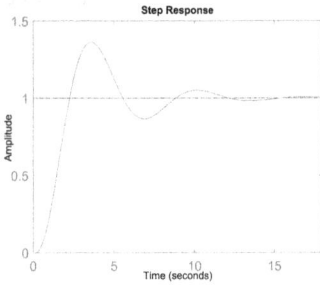

(a)

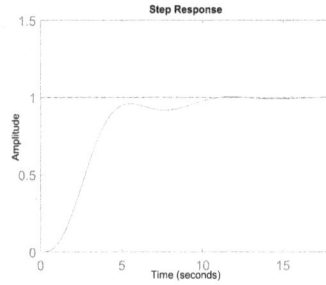

(b)

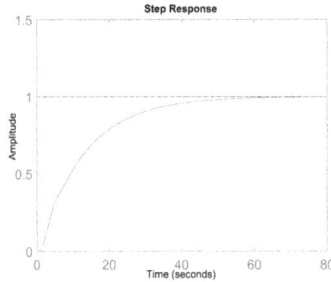

(c)

FIGURE 10.12

Step response for $G(s) = \frac{225p}{(s+p)(s^2+0.6s+1)(s^2+6s+225)}$ with: (a) $p = 4$; (b) $p = 0.4$; (c) $p = 0.08$.

10.8 MATLAB commands for the study of the step response and its characteristics

All parameters that have been presented in this lecture can be calculated in MATLAB®. The first option is using the command **step** to calculate and plot the step response of a system and, then, navigating in the figure by right-clicking on it and opening the menu that allows to select the character-istics to be calculated. These include: the peak response, the settling time, the rise time, and the steady state. Notice that the peak response includes information on the peak, the time of the peak, and the percentage overshoot. The settling time that is calculated in MATLAB® assumes $\varepsilon = 2\%$.

All these characteristics can be calculated using the command `stepinfo` as well. The command can be used either by inputting a transfer function or a set of step response data, e.g., two vectors, one containing time and the other the output values. An example of the first use of this command is the following:

```
s=tf('s')
wn=1;
xi=0.4;
mu=1;
G=mu*wn^2/(s^2+2*xi*wn*s+wn^2)
stepinfo(G)
```

Most of the results discussed in this lecture can be checked by calculating, with the command `step`, the step response in MATLAB® of the different systems under analysis. For instance, to study the effects of a negative zero, the following commands can be used

```
s=tf('s');
wn=1;
xi=0.3;
mu=1;
G0=mu*wn^2/(s^2+2*xi*wn*s+wn^2);
Gder=mu*wn^2/(s^2+2*xi*wn*s+wn^2)*s;
alpha=1;
G1=mu*wn^2*(s/(alpha*xi*wn)+1)/(s^2+2*xi*wn*s+wn^2);
alpha=2;
G2=mu*wn^2*(s/(alpha*xi*wn)+1)/(s^2+2*xi*wn*s+wn^2);
tfin=20;
figure,step(G1,tfin)
hold on
step(G2,tfin)
step(G0,tfin)
step(Gder,tfin)
```

Changing the sign of α, we can analyze the case of a positive zero. Con-versely, to study the effects of an additional pole, the following commands can be used

```
s=tf('s');
wn=1;
xi=0.3;
mu=1;
G0=mu*wn^2/(s^2+2*xi*wn*s+wn^2);
alpha=1;
G1=mu*wn^2/(s^2+2*xi*wn*s+wn^2)/(s/(alpha*xi*wn)+1);
alpha=2;
G2=mu*wn^2/(s^2+2*xi*wn*s+wn^2)/(s/(alpha*xi*wn)+1);
tfin=20;
figure,step(G1,tfin)
hold on
step(G2,tfin)
step(G0,tfin)
legend('\alpha=1','\alpha=2','\alpha=\infty')
```

Example 10.6 can be checked by using the following commands:

```
s=tf('s')
wn=1;
xi=0.3;
mu=1;
p=4;
G=mu*wn^2*p*15^2/(s^2+2*xi*wn*s+wn^2)/(s+p)/(s^2+2*0.2*15*s+15^2)
figure, step(G), ylim([0 1.5])
p=0.4;
G=mu*wn^2*p*15^2/(s^2+2*xi*wn*s+wn^2)/(s+p)/(s^2+2*0.2*15*s+15^2)
figure, step(G), ylim([0 1.5])
p=0.08
G=mu*wn^2*p*15^2/(s^2+2*xi*wn*s+wn^2)/(s+p)/(s^2+2*0.2*15*s+15^2)
figure, step(G), ylim([0 1.5])
```

10.9   Exercises

1. Using MATLAB®, calculate the step response for $G(s) = \frac{18}{s^2+0.6s+9}$ and evaluate $s_\%$, t_r, and $t_{s,2\%}$. Next compare the values obtained with available analytical expressions.

2. Using MATLAB®, calculate the step response for $G(s) = \frac{9(s+2)}{s^2+0.6s+9}$ and evaluate $s_\%$, t_r, and $t_{s,2\%}$. Next compare the values obtained with available analytical expressions.

3. Using MATLAB®, calculate the step response for $G(s) = \frac{9(2-s)}{s^2+0.6s+9}$ and evaluate $s_\%$, t_r, and $t_{s,2\%}$. Next compare the values obtained with available analytical expressions.

4. Given the third-order system $G(s) = \frac{12}{s^3+2s^2+3.36s+0.6}$, evaluate if it is more appropriate a first-order or a second-order dominant pole approximation.

5. Given the third-order system $G(s) = \frac{12s+2.28}{0.19s^3+0.38s^2+0.6384s+0.114}$, evaluate if it is more appropriate a first-order or a second-order dominant pole approximation.

11

Lecture 11—Frequency response and Bode diagrams

In this lecture, an important property of continuous time LTI system is studied. This is the frequency response, namely the response of the system when the input is a sinusoidal signal. This notion is fundamental as it paves the way to the interpretation of dynamical systems as filters, an aspect that is also illustrated in this lecture.

11.1 Frequency response

There is an important result for BIBO stable systems concerning the response obtained when the system is forced with a sinusoidal input.

Given a BIBO stable system with transfer function $G(s)$, driven by an input $u(t) = U \sin(\omega t + \varphi_u)$, then the steady-state output of the system is given by

$$y(t) = Y \sin(\omega t + \varphi_y) \tag{11.1}$$

where

$$Y = |G(j\omega)|U \tag{11.2}$$

and

$$\varphi_y = \varphi_u + \angle G(j\omega) \tag{11.3}$$

After a transient phase, whose duration depends on the system poles, then the output of the system looks like a sinusoidal waveform. The frequency of this signal is the same as the input signal, while its amplitude and phase are given by Eqs. (11.2) and (11.3), respectively.

In fact, since

$$\sin \theta = \frac{e^{j\theta} - e^{-j\theta}}{2j} \tag{11.4}$$

DOI: 10.1201/9781003487289-11

we have that

$$\mathcal{L}(\sin \omega t) = \mathcal{L}\left(\frac{e^{j\theta} - e^{-j\theta}}{2j}\right) = \frac{1}{2j(s - j\omega)} - \frac{1}{2j(s + j\omega)} \tag{11.5}$$

Using this result, we find that the system output is as follows:

$$Y(s) = G(s)U(s) = G(s)\left(\frac{1}{2j(s - j\omega)} - \frac{1}{2j(s + j\omega)}\right) \tag{11.6}$$

By expanding in simple fractions, we get

$$Y(s) = \frac{R_1}{s - p_1} + \ldots + \frac{R_n}{s - p_n} + \frac{R_{s,1}}{s - j\omega} + \frac{R_{s,2}}{s + j\omega} \tag{11.7}$$

We notice that the first terms, namely those associated with the system poles, once the inverse Laplace transform is applied, lead to time functions that go to zero as $t \to +\infty$. So, the steady-state output is given by the remaining two terms:

$$Y_{ss}(s) = \frac{R_{s,1}}{s - j\omega} + \frac{R_{s,2}}{s + j\omega} \tag{11.8}$$

To calculate it, we need to calculate $R_{s,1}$ and $R_{s,2}$. They are given by

$$R_{s,1} = \lim_{s \to j\omega} (s - j\omega)Y(s) = \lim_{s \to j\omega} \frac{G(s)}{2j} = \frac{G(j\omega)}{2j} \tag{11.9}$$

and

$$R_{s,2} = \lim_{s \to -j\omega} (s + j\omega)Y(s) = \lim_{s \to -j\omega} -\frac{G(s)}{2j} = -\frac{G(-j\omega)}{2j} \tag{11.10}$$

Replacing $R_{s,1}$ and $R_{s,2}$ in (11.8), we obtain

$$Y_{ss}(s) = \frac{G(j\omega)}{2j}\frac{1}{(s - j\omega)} - \frac{G(-j\omega)}{2j}\frac{1}{(s + j\omega)} \tag{11.11}$$

and so

$$y_{ss}(t) = \frac{G(j\omega)}{2j}e^{j\omega t} - \frac{G(-j\omega)}{2j}e^{-j\omega t} \tag{11.12}$$

Since $G(j\omega) = |G(j\omega)|e^{j\angle G(j\omega)}$ and $G(-j\omega) = |G(j\omega)|e^{-j\angle G(j\omega)}$, we get

$$y_{ss}(t) = \frac{|G(j\omega)|}{2j}e^{j\omega t + j\angle G(j\omega)} - \frac{|G(j\omega)|}{2j}e^{-j\omega t - j\angle G(j\omega)} \tag{11.13}$$

and finally

$$y_{ss}(t) = |G(j\omega)|\frac{e^{j\omega t + j\angle G(j\omega)} - e^{-j\omega t - j\angle G(j\omega)}}{2j} = |G(j\omega)|\sin(\omega t + \angle G(j\omega)) \tag{11.14}$$

The previous result can be extended to periodic signals other than the sinusoidal and cosinusoidal waveforms. In fact, consider a periodic signal $u(t)$ of period T. It can be written as a Fourier series:

$$u(t) = a_0 + \sum_{k=1}^{\infty} a_k \cos(k\omega t + \varphi_k) \qquad (11.15)$$

where $\omega = \frac{2\pi}{T}$. By the superposition principle, we find that, for a BIBO stable system, the steady-state output in response to this input is

$$y_{ss}(t) = G(0)a_0 + \sum_{k=1}^{\infty} a_k |G(jk\omega t)| \cos(k\omega t + \varphi_k + \angle G(jk\omega t)) \qquad (11.16)$$

From these considerations, we realize that $|G(j\omega)|$ represents the amplification/attenuation factor of the component at a frequency equal to ω of the spectrum of the input signal. Correspondingly, $\angle G(j\omega)$ is the quantity by which the component will be shifted in phase. Having a graphical representation of these two quantities is thus extremely useful. The plots of $|G(j\omega)|$ and $\angle G(j\omega)$ are called the Bode diagrams and are dealt with in the next section. Although the frequency response interpretation of the Bode diagrams can be given only for BIBO stable systems, the Bode diagrams are a more general and useful mathematical tool, and, therefore, we will develop them for both stable and unstable systems.

11.2   Bode diagrams

Bode diagrams are drawn starting from the Bode form as in (7.36), here recalled for convenience:

$$G(s) = \frac{\mu \Pi_i (s/z_i + 1) \Pi_i (s^2/\omega_{z,i}^2 + 2\zeta_i s/\omega_{z,i} + 1)}{s^g \Pi_i (s/p_i + 1) \Pi_i (s^2/\omega_{n,i}^2 + 2\xi_i s/\omega_{n,i} + 1)} \qquad (11.17)$$

that for $s = j\omega$ becomes

$$G(j\omega) = \frac{\mu \Pi_i (j\omega/z_i + 1) \Pi_i (-\omega^2/\omega_{z,i}^2 + 2j\zeta_i \omega/\omega_{z,i} + 1)}{s^g \Pi_i (j\omega/p_i + 1) \Pi_i (-\omega^2/\omega_{n,i}^2 + 2j\xi_i \omega/\omega_{n,i} + 1)} \qquad (11.18)$$

Since, in general, ω spans several orders of magnitude, and Bode diagrams use a logarithmic scale for the x-axis.

For the Bode diagram of the magnitude, the value of $|G(j\omega)|$ is expressed in decibel (dB):

$$|G(j\omega)|_{dB} = 20 \log |G(j\omega)| \tag{11.19}$$

The magnitude of $G(j\omega)$ in (11.18) can be written as follows:

$$
\begin{aligned}
|G(j\omega)|_{dB} = \quad & 20 \log |\mu| - 20g \log |j\omega| + \sum_i 20 \log |j\omega/z_i + 1| \\
& - \sum_i 20 \log |j\omega/p_i + 1| \\
& + \sum_i 20 \log | - \omega^2/\omega_{z,i}^2 + 2j\zeta_i\omega/\omega_{z,i} + 1| \\
& - \sum_i 20 \log | - \omega^2/\omega_{n,i}^2 + 2j\xi_i\omega/\omega_{n,i} + 1|
\end{aligned}
\tag{11.20}
$$

From this expression we conclude that the Bode diagram of the amplitude can be drawn by, first, drawing the diagrams of the single terms, and then summing them.

The phase is represented either in degrees or radians. Here we will use degrees.

For the phase we have

$$
\begin{aligned}
\angle G(j\omega) = \angle\mu - g\angle j\omega + \sum_i \angle(j\omega/z_i + 1) - \sum_i \angle(j\omega/p_i + 1)+ \\
+ \sum_i \angle(-\omega^2/\omega_{z,i}^2 + 2j\zeta_i\omega/\omega_{z,i} + 1) - \sum_i \angle(-\omega^2/\omega_{n,i}^2 + 2j\xi_i\omega/\omega_{n,i} + 1)
\end{aligned}
\tag{11.21}
$$

This yields that, also for the phase, the diagram can be obtained as the sum of elementary terms.

Since for any complex number c, $20 \log |\frac{1}{c}| = -20 \log |c|$ and $\angle\frac{1}{c} = -\angle c$, we can concentrate on studying the Bode diagrams of the following elementary terms:

$$G(s) = \mu \tag{11.22}$$

$$G(s) = \frac{1}{s} \tag{11.23}$$

$$G(s) = \frac{1}{s/p + 1} \tag{11.24}$$

$$G(s) = \frac{1}{s^2/\omega_n^2 + 2\xi s/\omega_n + 1} \tag{11.25}$$

11.2.1   Bode diagram of $G(s) = \mu$

The Bode diagram of the gain μ does not depend on ω, so it is a constant value. For the amplitude, this constant value is $20 \log |\mu|$, and is positive if $|\mu| > 1$, and negative if $|\mu| < 1$. For the phase, the constant value is $0°$ if $\mu > 0$, as a real positive number has zero phase, while it is $-180°$ if $\mu < 0$, as a real negative number has a phase equal to $-\pi$ (or π). As an example, Fig. 11.1 illustrates the Bode diagram for $\mu = 5$.

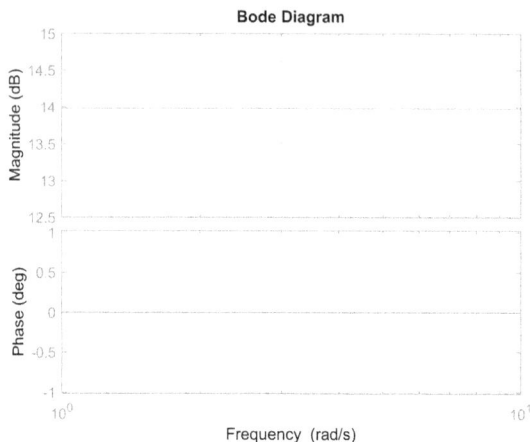

FIGURE 11.1
Bode diagram of $G(s) = \mu$ with $\mu = 5$.

11.2.2   Bode diagram of $G(s) = \frac{1}{s}$

We now study the Bode diagram of a transfer function with a single pole at the origin, namely $G(s) = \frac{1}{s}$, for which $G(j\omega) = \frac{1}{j\omega}$. For the amplitude we have that $20 \log |G(j\omega)| = -20 \log \omega$. Taking into account that in the x-axis we represent $\log \omega$, the Bode diagram of the amplitude is a straight line of slope equal to $-20dB$ per decade. This line crosses the x-axis at $\omega = 1$, as $-20 \log 1 = 0$.

For the phase, we notice that, since $\omega > 0$, the complex number $\frac{1}{j\omega}$ always lies on the negative imaginary axis. Since the phase of a complex number with zero real part and negative imaginary part is $-90°$, the phase of $G(j\omega) = \frac{1}{j\omega}$ is a constant value at $-90°$.

The Bode diagram of $G(s) = \frac{1}{s}$ is shown in Fig. 11.2.

11.2.3   Bode diagram of $G(s) = \frac{1}{s/p+1}$

We consider now a system with a single pole in $s = -p$, namely $G(s) = \frac{1}{s/p+1}$. Replacing $s = j\omega$, we have $G(j\omega) = \frac{1}{j\omega/p+1}$. The amplitude of $G(j\omega)$ is given by:

$$|G(j\omega)| = \frac{1}{\sqrt{\left(\frac{\omega^2}{p^2} + 1\right)}} \tag{11.26}$$

For $\omega/p \ll 1$, $|G(j\omega)| \simeq 1$, and so $|G(j\omega)|_{dB} \simeq 0$. For $\omega/p \gg 1$, instead $|G(j\omega)| \simeq \frac{p}{\omega}$, and so $|G(j\omega)|_{dB} \simeq -20 \log |\omega/p|$. This suggests to approximate

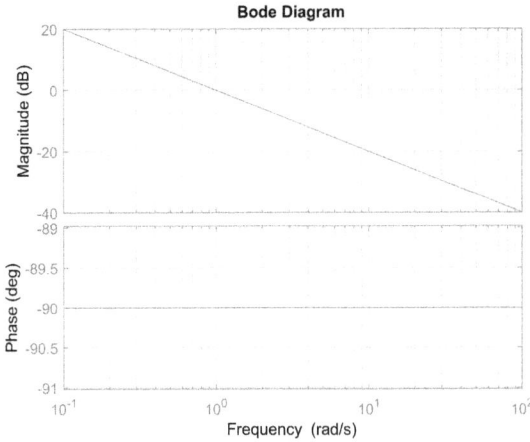

FIGURE 11.2
Bode diagram of $G(s) = \frac{1}{s}$.

the Bode diagram of $G(s) = \frac{1}{s/p+1}$ with two traits, one flat up to $\omega = p$, and the other one with a negative slope of -20dB per decade. The largest discrepancy between real and approximate Bode diagram occurs at $\omega = p$, where $|G(jp)| = \frac{1}{\sqrt{2}}$, that is, $|G(jp)|_{dB} = -3dB$.

The phase of $G(j\omega) = \frac{1}{j\omega/p+1}$ is given by:

$$\angle G(j\omega) = \frac{1}{j\omega/p + 1} = -\arctan\frac{\omega}{p} \qquad (11.27)$$

To approximate this function, one can use a piece wise linear function with three traits, considering that at $\omega = p$ $\angle G(j\omega) = -45°$, that $\angle G(j\omega) \simeq 0$ for $\omega/p \ll 0.1$, and that $\angle G(j\omega) \simeq -90°$ for $\omega/p \gg 10$.

Fig. 11.3 shows the Bode diagram of a transfer function with a single negative pole, namely $G(j\omega) = \frac{1}{j\omega/p+1}$. Here $p = 1$. Alternatively, one can imagine that the frequency axis is normalized by p, namely, the frequency is expressed as ω/p.

So far we have considered $p > 0$, so the pole is negative. When $p < 0$ (positive pole), because of Eq. (11.26) the magnitude remains the same, while, due to Eq. (11.27), the phase changes its sign. The overall result is that the magnitude diagram remains the same, while the phase diagram is mirrored with respect to the y-axis. Fig. 11.4 shows an example with $p = -1$.

Following similar arguments, the Bode diagram of a transfer function with a single zero, either negative or positive, can be derived, namely $G(s) = s/z+1$ with $z > 0$ or $z < 0$. Moving from $G(s) = \frac{1}{s/p+1}$ to $G(s) = s/z + 1$, the same

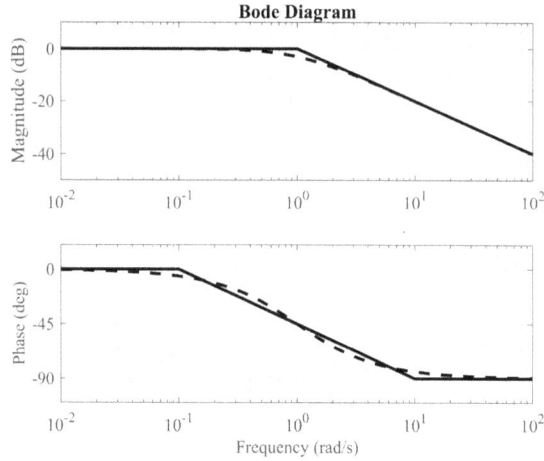

FIGURE 11.3
Bode diagram of $G(s) = \frac{1}{s/p+1}$ with $p = 1$.

types of approximation can be used, but both the magnitude and the phase change their sign. As an example, panels (a) and (b) of Fig. 11.5 show the Bode diagram of $G(s) = s/z + 1$ with $z = 1$ and $z = -1$, respectively.

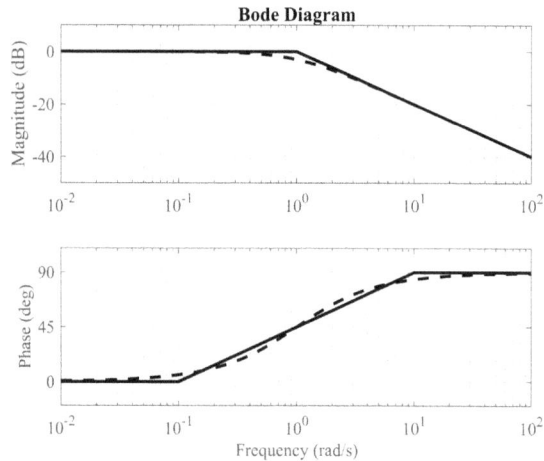

FIGURE 11.4
Bode diagram of $G(s) = \frac{1}{s/p+1}$ with $p = -1$.

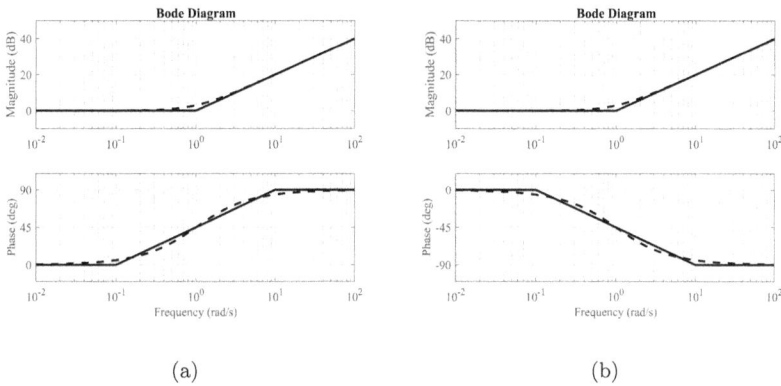

(a)   (b)

FIGURE 11.5
Bode diagram of $G(s) = s/z + 1$ with (a) $z = 1$, and (b) $z = -1$.

11.2.4   Bode diagram of $G(s) = \frac{1}{s^2/\omega_n^2 + 2\xi s/\omega_n + 1}$

We consider now a system with a pair of complex and conjugate poles, that is $G(s) = \frac{1}{s^2/\omega_n^2 + 2\xi s/\omega_n + 1}$ with $\omega_n > 0$ and $\xi > 0$. Replacing s with $j\omega$, we get $G(j\omega) = \frac{1}{-\omega^2/\omega_n^2 + 2j\xi_i \omega/\omega_n + 1}$. The amplitude of $G(j\omega)$ reads: $|G(j\omega)| = \frac{1}{\sqrt{(1-\omega^2/\omega_n^2)^2 + 4\xi_i^2 \omega^2/\omega_n^2}}$. For $\omega \ll \omega_n$ we find that $|G(j\omega)| \simeq 1$, while for $\omega \gg \omega_n$ we find that $|G(j\omega)| \simeq \frac{\omega_n^2}{\omega^2}$. It follows that: for $\omega \ll \omega_n$, $|G(j\omega)|_{dB} \simeq 0$, and, for $\omega \gg \omega_n$, $|G(j\omega)|_{dB} \simeq -40 \log |\omega/\omega_n|$. According to this approximation the Bode diagram of the magnitude is flat up to $\omega/\omega_n = 1$, and then it decreases with a slope of $-40dB$ per decade. For $\xi < 1/\sqrt{2}$, the diagram has a peak (see also Fig. 11.6) at the so-called *resonance frequency* ω_r:

$$\omega_r = \omega_n \sqrt{1 - 2\xi^2} \tag{11.28}$$

The magnitude peak is equal to

$$|G(j\omega_r)|_{dB} = \frac{1}{2|\xi|\sqrt{1-\xi^2}} \tag{11.29}$$

Instead, at the frequency $\omega = \omega_n$ the magnitude is equal to

$$|G(j\omega_n)|_{dB} = \frac{1}{2|\xi|} \tag{11.30}$$

Notice that, for $\xi = 0$, $\omega_r = \omega_n$, and the resonance peak is infinite. For larger values of $\xi > 0$, ω_r becomes increasingly different from ω_n. The magnitude diagram of Fig. 11.6 should be considered to correct the error made by using the approximation introduced above. This becomes increasingly important for smaller values of ξ.

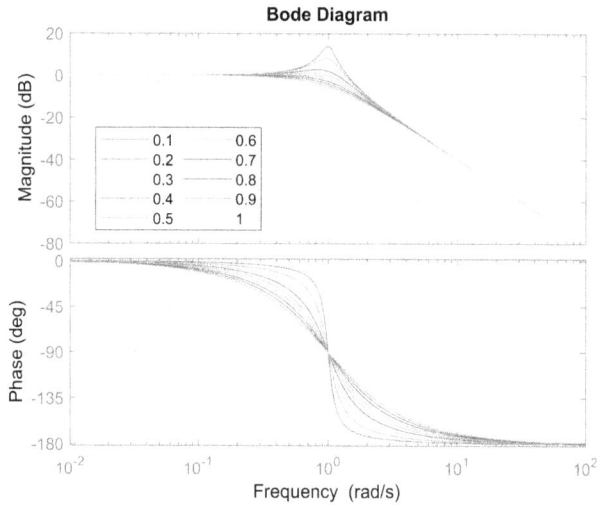

FIGURE 11.6
Bode diagram of $G(s) = \frac{1}{s^2/\omega_n^2 + 2\xi s/\omega_n + 1}$ with $\omega_n = 1$ for several values of ξ.

Another important characteristic of the frequency response of this system is the bandwidth, defined as the value of ω for which $|G(j\omega)| = 1/\sqrt{2}$. This is indicated as B_3, with reference to the fact that $20\log(1/\sqrt{2}) = -3$dB. From the condition $|G(j\omega)| = 1/\sqrt{2}$, one gets that

$$B_3 = \omega_n \sqrt{(1 - 2\xi^2) + \sqrt{4\xi^4 - 4\xi^2 + 2}} \tag{11.31}$$

Let us now study the phase of $G(j\omega) = \frac{1}{-\omega^2/\omega_n^2 + 2j\xi_i\omega/\omega_n + 1}$. For $\xi \neq 0$, we have that

$$\angle G(j\omega) = \begin{cases} -\arctan\frac{2\xi_i\omega/\omega_n}{1-\omega^2/\omega_n^2}, & \text{if } \omega < \omega_n \\ -\arctan\frac{2\xi_i\omega/\omega_n}{1-\omega^2/\omega_n^2} - \pi, & \text{if } \omega > \omega_n \end{cases} \tag{11.32}$$

For $\xi = 0$, $G(j\omega) = \frac{1}{-\omega^2/\omega_n^2 + 1}$ is always a real number; it is positive if $\omega < \omega_n$, and negative if $\omega > \omega_n$. Hence, we have that

$$\angle G(j\omega) = \begin{cases} 0, & \text{if } \omega < \omega_n \\ -\pi, & \text{if } \omega > \omega_n \end{cases} \tag{11.33}$$

So far, we have considered the case of complex conjugate poles with negative (or zero) real part. If $-1 < \xi < 0$, then the poles of $G(s)$ are complex and conjugate with a positive real part. The Bode diagram of the magnitude remains the same, while, since the phase has the opposite sign, the diagram related to the phase is reversed.

Lastly, for complex and conjugate zeros, namely $G(s) = s^2/\omega_n^2 + 2\xi s/\omega_n + 1$, both the magnitude and phase diagrams are reversed.

11.3 Examples of Bode diagrams

In this section we illustrate through several examples the procedure to draw the Bode diagram of a system. To plot the Bode diagram in MATLAB®, we use either the command **bode** and the function **asbode** developed by Alessandro Giua and Carla Seatzu and publicly available at https://www.alessandro-giua.it/UNICA/ASD/materiale.html. This latter function also generates the asymptotic diagrams, making it particularly useful for educational purposes.

Example 11.1 _____

Bode diagram of $G(s) = \frac{10}{s(s+5)}$.

First of all, we write $G(s)$ in Bode form: $G(s) = \frac{2}{s(s/5+1)}$, which allows one to identify the three terms to consider to draw the Bode diagram of the system:

- a positive gain $\mu = 2$ $(20\log(\mu) = 6dB)$;
- a pole in $s = 0$;
- a negative pole in $s = -5$.

The asymptotic Bode diagrams of these three contributions, along with their sum, are shown in Fig. 11.7(a), obtained using the MATLAB® command asbode(10,[1 5 0]), while the overall Bode diagram, obtained with the command bode(10,[1 5 0]), is shown in Fig. 11.7(b).

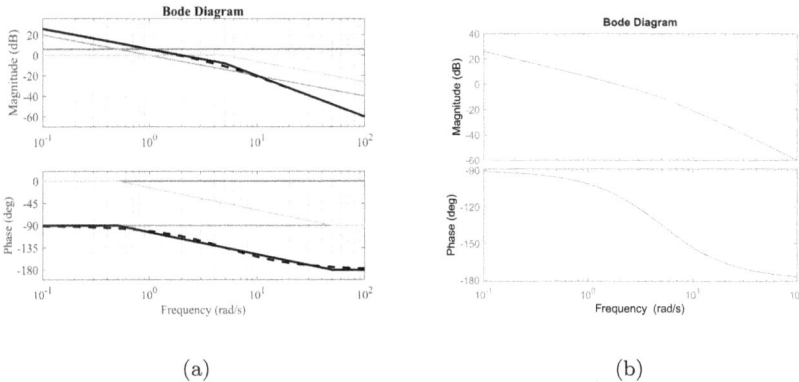

(a)   (b)

FIGURE 11.7
Bode diagram of $G(s) = \frac{10}{s(s+5)}$: (a) approximated; (b) numerical.

Example 11.2

Bode diagram of $G(s) = \frac{4000}{s^3+4s^2+400s}$.

The first step is to write $G(s)$ in Bode form: $G(s) = \frac{10}{s(s^2/400+0.01s+1)}$. To draw the Bode diagram of the system, we have to take into account three terms:

- a positive gain $\mu = 10$ $(20\log(\mu) = 20dB)$;

- a pole in $s = 0$;

- a pair of complex and conjugates poles with negative real part, having $\xi = 0.1$, and $\omega_n = 20$.

The asymptotic Bode diagrams of these three terms, along with their sum, are shown in Fig. 11.8(a), obtained using the MATLAB® command asbode(4000,[1 4 400 0]), while the overall Bode diagram, obtained with the command bode(4000,[1 4 400 0]), is shown in Fig. 11.8(b).

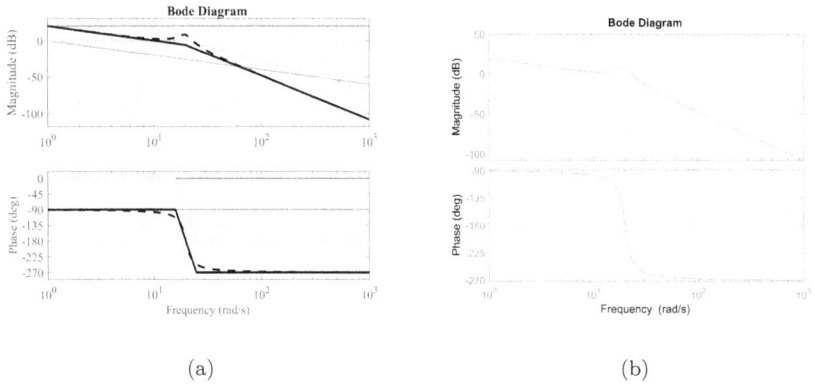

(a)   (b)

FIGURE 11.8
Bode diagram of $G(s) = \frac{4000}{s^3+4s^2+400s}$: (a) approximated; (b) numerical.

Example 11.3

Bode diagram of $G(s) = \frac{s+3}{s^2+11s+10}$.

The transfer function $G(s)$ in Bode form reads: $G(s) = \frac{0.3(s/3+1)}{(s+1)(s/10+1)}$. To draw the Bode diagram of the system, we have to take into account four terms:

- a positive gain $\mu = 0.3$ $(20\log(\mu) = -10dB)$;

- a negative zero in $s = -3$;

- a negative pole in $s = -1$;

- a negative pole in $s = -10$.

The asymptotic Bode diagrams of these four terms, along with their sum, are shown in Fig. 11.9(a), obtained using the MATLAB® command asbode([1 3],[1 11 10]), while the overall Bode diagram, obtained with the command bode([1 3],[1 11 10]), is shown in Fig. 11.9(b).

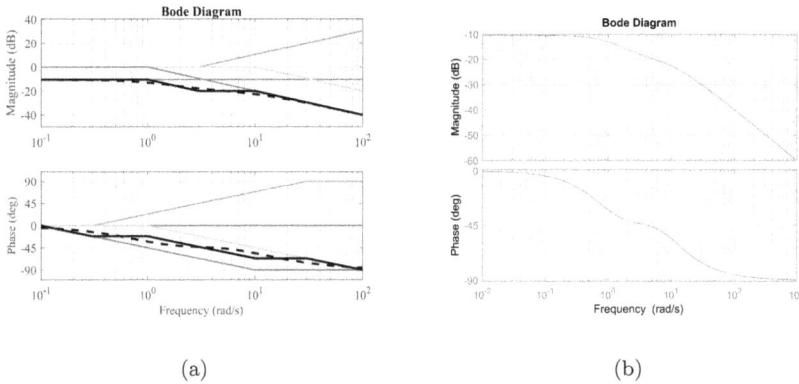

(a)   (b)

FIGURE 11.9
Bode diagram of $G(s) = \frac{s+3}{s^2+11s+10}$: (a) approximated; (b) numerical.

Example 11.4 _____

Bode diagram of $G(s) = \frac{s+3}{s^2+9s-10}$.

The transfer function $G(s)$ in Bode form reads: $G(s) = \frac{-0.3(s/3+1)}{(1-s)(s/10+1)}$. We have

- a negative gain $\mu = -0.3$ $(20\log(|\mu|) = -10dB)$;

- a negative zero in $s = -3$;

- a positive pole in $s = 1$;

- a negative pole in $s = -10$.

With respect to the previous example, the gain is negative, rather than positive, and one of the two poles is positive rather than negative. These differences do not impact on the Bode diagram of the amplitude, which is the same as in the previous example. However, the Bode diagram of the phase looks different, as the negative gain contributes with $-180°$, and the positive pole with a phase ranging from $0°$ to $90°$. The asymptotic Bode diagrams, obtained using the MATLAB® command asbode([1 3],[1 9 -10]), are shown in Fig. 11.10(a), while the overall Bode diagram, obtained with the command asbode([1 3],[1 9 -10]), is shown in Fig. 11.10(b).

11.4 Filter characteristics of LTI systems

In this section, we discuss how LTI systems may act as filters for the input signals, namely how they produce an output that is function of the input with selective frequency characteristics. We illustrate this property of LTI systems with some examples.

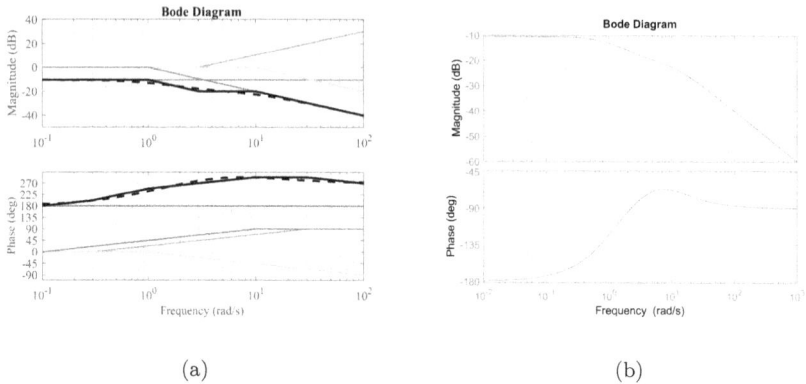

(a)   (b)

FIGURE 11.10
Bode diagram of $G(s) = \frac{s+3}{s^2+9s-10}$: (a) approximated; (b) numerical.

Consider the following four LTI systems:

$$G_1(s) = \frac{1}{0.1s + 1} \tag{11.34}$$

$$G_2(s) = \frac{0.1s}{0.1s + 1} \tag{11.35}$$

$$G_3(s) = \frac{s}{(0.01s + 1)(s + 1)} \tag{11.36}$$

$$G_4(s) = \frac{(0.1s + 1)(s + 1)}{(0.01s + 1)(10s + 1)} \tag{11.37}$$

Their Bode diagrams are shown in Fig. 11.11. All systems are BIBO stable, so the frequency response theorem applies. Consider first $G_1(s)$ (Fig. 11.11(a)) and a sinusoidal input with $\omega < 10$ rad/s. Since the magnitude is close to one in that frequency range, the system output is almost the same as the input signal. Conversely, a signal with $\omega > 10$ rad/s is attenuated at the output of the system. We conclude that the system behaves as a *low-pass filter*: in fact, signals with frequencies lower than 10 rad/s pass almost unaltered through the system, while signal with frequencies higher than 10 rad/s are attenuated at the output of the system.

System $G_2(s)$ (Fig. 11.11(b)) displays a different behavior. In fact, signals with frequencies lower than 10 rad/s are now attenuated, while those with frequencies higher than 10 rad/s pass almost unaltered. This behavior is characteristic of *high-pass filters*.

For the system $G_3(s)$ a frequency band, from 1 rad/s to 100 rad/s, can be defined. Signals with frequencies within this band pass through the system, while other frequencies are attenuated at the output. Lastly, for the system $G_4(s)$ a frequency band, from 0.1 rad/s to 100 rad/s, can be defined, but this time signals with frequencies within this band are attenuated, while those with frequencies outside this band pass through the system. The behavior of systems $G_3(s)$ and $G_4(s)$ is characteristic of a *band-pass* and a *band-stop filter*, respectively.

In summary, LTI systems can behave as:

- low-pass filters that pass low frequencies and block high frequencies;

- high-pass filters that pass high frequencies and block low frequencies;

- band-pass filters that pass a band of frequencies;

- band-stop filters that pass high and low frequencies outside of a given band.

There are many mathematical procedures that, starting from specified frequency characteristics, allow for the design of a filter, potentially including other design constraints. Although this book will not delve into these procedures, the reader should know that this task can be accomplished using various techniques.

11.5 MATLAB examples

In this section, we discuss two MATLAB® examples illustrating the frequency response of a system (MATLAB® exercise 11.1), and the filtering effects of dynamical systems (MATLAB® exercise 11.2).

MATLAB® exercise 11.1 _____

Let us consider the system $G(s) = 1/(s+1)$ with input $u(t) = 3\sin(2t + \pi/4)$. Since the system is BIBO stable, the steady-state output is also sinusoidal with amplitude and phase given by Eq. (11.2) and Eq. (11.3), respectively. Let us now check this result by using MATLAB® to calculate the output.
First, we define the system $G(s)$ with the commands:

```
s=tf('s')
G=1/(s+1)
```

Next, we calculate the system response to $u(t) = 3\sin(2t + \pi/4)$ and plot it along with the input:

```
t=[0:0.01:20];
u=3*sin(2*t+pi/4);
y=lsim(G,u,t);
figure,plot(t,u,t,y)
```

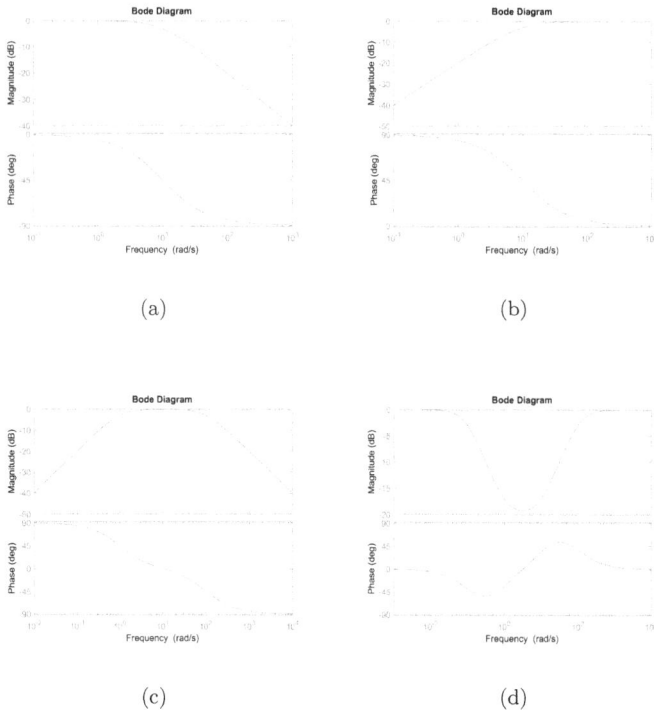

(a)

(b)

(c)

(d)

FIGURE 11.11

Bode diagrams of: (a) $G_1(s) = \frac{1}{0.1s+1}$; (b) $G_2(s) = \frac{0.1s}{0.1s+1}$; (c) $G_3(s) = \frac{s}{(0.01s+1)(s+1)}$; (d) $G_4(s) = \frac{(0.1s+1)(s+1)}{(0.01s+1)(10s+1)}$.

Fig. 11.12 shows the result. From this figure, we find that, after a quick transient, the output converges to a sinusoidal waveform with amplitude equal to $Y = 1.337$. Hence, we have that $Y/U = 1.342/3 = 0.4473$. This value is in agreement with Eq. (11.2) with $\omega = 2$, as $|G(j2)| = 1/\sqrt{5} = 0.4472$. From the time evolution of $u(t)$ and $y(t)$, we also calculate the phase lag between the two signals by considering the times of two consecutive peaks of the signals. For instance, we can focus on the peak of the input at $t = 16.1$ and find that the next peak of the output occurs at $t = 16.66$ s. This yields a time difference equal to $\Delta T = 16.1 - 16.66 = -0.5600$ s. Since the period of the signals is $T = 2\pi/\omega = \pi$, by calculating $\Delta T/T \cdot 360 = -0.56 \cdot 360/\pi = -64.1713°$ we get the phase lag in degrees. This matches Eq. (11.3), from which we get $\langle G(j2) = -\arctan(2) \cdot 180/\pi = -63.4349°$.

Notice that magnitude and phase can also be calculated using the MATLAB® command bode with a single value of frequency $\omega = 2$:

```
[magnitude,phase]=bode(G,2)
```

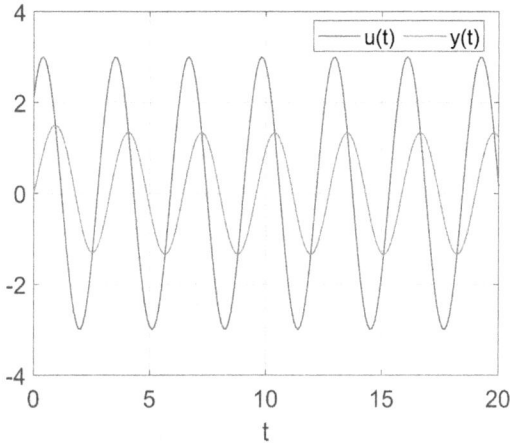

FIGURE 11.12
Response $y(t)$ of the system $G(s) = 1/(s+1)$ to the input $u(t) = 3\sin(2t+\pi/4)$.

MATLAB® exercise 11.2 _____

Let us now consider the system $G(s) = \frac{1000}{(s+10)^3}$. First, we define the system in MATLAB® and draw its Bode diagram with the commands:

```
s=tf('s')
G=1000/(s+10)^3
figure,bode(G)
```

Then, let us calculate the response to several signals. We begin with $u = 2\sin(3t)$. From the Bode diagram, we find that at $\omega = 3$ rad/s the magnitude is close to 0dB. Hence, the amplitude of this signal at the output will be slightly less than the input. To check this, we use the following commands:

```
t=[0:0.001:10];
u=2*sin(3*t);
y=lsim(G,u,t);
figure,plot(t,u,t,y)
```

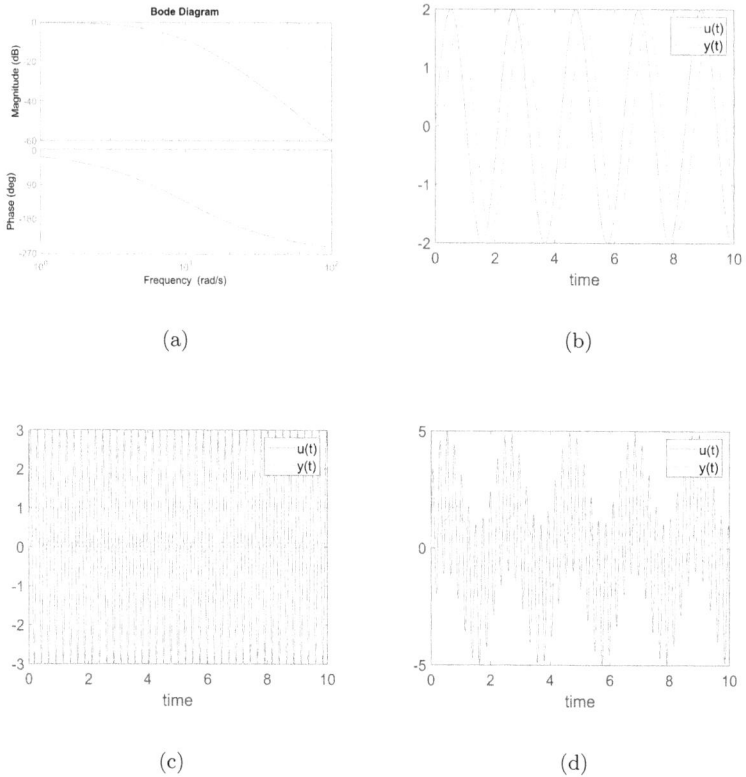

(a)

(b)

(c)

(d)

FIGURE 11.13

An example of linear distortion. (a) Bode diagram of $G(s) = \frac{1000}{(s+10)^3}$. (b) System response to $u = 2\sin(3t)$. (c) System response to $u = 3\sin(26t)$. (d) System response to $u = 2\sin(3t) + 3\sin(26t)$.

For higher frequencies, the amplitude will decrease progressively. For instance, let us consider $\omega = 26$ rad/s:

```
u=3*sin(26*t);
y=lsim(G,u,t);
figure,plot(t,u,t,y)
```

Let us now consider a signal that is obtained as the sum of the two terms previously analyzed and calculate the corresponding output:

```
u=2*sin(3*t)+3*sin(26*t);
y=lsim(G,u,t);
figure,plot(t,u,t,y)
```

The output is noticeably distorted due to the differing amplification factors of its two components, illustrating an example of linear distortion.

11.6 Exercises

1. Draw the Bode diagram of $G(s) = \frac{s-2}{s(s^2+36)}$.

2. Draw the Bode diagram of $G(s) = \frac{10}{s^2(2-s)}$.

3. Draw the Bode diagram of $G(s) = \frac{s}{(s+1)(20-s)}$.

4. Draw the Bode diagram of $G(s) = \frac{s+0.1}{s(s^2+2s+25)}$.

5. Draw the Bode diagram of $G(s) = \frac{s^2+9}{s(s-0.2)(s+30)}$.

12

Lecture 12—Closed-loop stability

In this lecture, we analyze closed-loop stability based on the characteristics of the open-loop system. We will begin with step-by-step examples on how to construct the Nyquist plot of a transfer function. Following this, we will introduce the fundamental Nyquist stability criterion and apply it to a variety of case studies.

12.1 Stability under nominal conditions

Closed-loop stability is the primary requirement that a control system must satisfy, and the controller design must prioritize this fundamental aspect. Here, we examine how closed-loop stability can be analyzed based on the characteristics of the open-loop system. To this aim, we refer to the negative feedback control scheme of Fig. 12.1, where we assume that there are no simplifications between any positive zero and pole of $L(s)$. The closed-loop transfer function is given by

$$W(s) = \frac{L(s)}{1 + L(s)} \tag{12.1}$$

that shows that the closed-loop stability is determined by the following equation:

$$1 + L(s) = 0 \tag{12.2}$$

In fact, the roots of $1 + L(s) = 0$ are the closed-loop poles. The stability of $W(s)$ can be studied with some of the methods that we have already introduced in the previous lectures, e.g., the Routh criterion. However, there are cases where such criteria cannot be applied (e.g., in the presence of a time-delay) or are not adequate, as the task is the synthesis of the controller rather than the analysis of the stability of a system with given parameters. For this reason an useful criterion, which allows to study the closed-loop stability from the characteristics of the open-loop transfer function, namely $L(s)$, is the Nyquist stability criterion, which is based on the Nyquist diagram, discussed in the next sections.

DOI: 10.1201/9781003487289-12

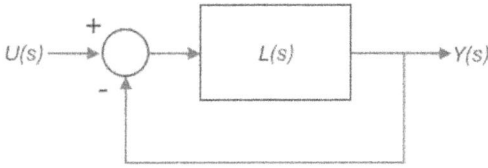

FIGURE 12.1
Reference scheme used for the study of the closed-loop stability.

12.2   The Nyquist diagram

To define the Nyquist diagram, we need first to introduce the notion of Nyquist path. The Nyquist path is a closed path that encircles all poles and zeros of $L(s)$ that lie in the open right half plane. If $L(s)$ has no poles on the imaginary axis, the Nyquist path is composed of two segments: one is the imaginary axis from $-j\infty$ to $+j\infty$, and the other is a semicircle of infinite radius encircling the right half plane, as shown in Fig. 12.2(a) and in Fig. 12.2(b). In the first case (Fig. 12.2(a)), $L(s)$ has no poles in the open right half plane, while in the second case (Fig. 12.2(b)) $L(s)$ has two complex and conjugate poles with positive real part. It is important to distinguish between these two cases to accurately determine the closed-loop stability.

If $L(s)$ has one or more poles on the imaginary axis, these have to be excluded to avoid ambiguity in counting how many poles are encircled by the

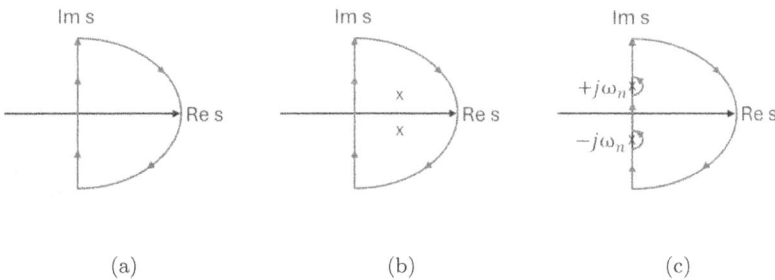

(a)   (b)   (c)

FIGURE 12.2
Examples of Nyquist path: (a) $L(s)$ with no poles on the imaginary axis and no poles in the open right half plane; (b) $L(s)$ with no poles on the imaginary axis and two poles in the open right half plane; (c) $L(s)$ having a pair of imaginary poles in $s_{1,2} = \pm j\omega_n$.

Nyquist plot. To this aim, as shown in Fig. 12.2(c), in the proximity of those poles it is possible to consider a semicircle of small radius ε, such that rather than replacing s with $j\omega$ it is replaced by $s = \pm j\omega_n + \varepsilon e^{j\theta}$ with $\theta \in [-\pi/2, \pi/2]$ and $\varepsilon \to 0$.

The Nyquist diagram is the mapping of the Nyquist path through the function $L(s)$. It is a closed curve in the complex plane obtained by replacing $s = j\omega$ in $L(s)$ and calculating the real and imaginary part of $L(j\omega)$ for ω taking values from $-\infty$ to $+\infty$. This corresponds to map the segment of the Nyquist path that lies on the imaginary axis. All the analysis of this lecture is carried out under the hypothesis that $\lim_{s \to \infty} L(s) = 0$; this is in fact a strict requirement to apply the criterion for the closed-loop stability that is introduced in the next section. Under this assumption, which holds for all $L(s)$ that are strictly proper, the portion of the Nyquist path that corresponds to the semicircle of infinite radius is mapped into a single point on the Nyquist diagram, namely the origin.

Since $L(-j\omega) = \bar{L}(j\omega)$, then the Nyquist diagram can be viewed as formed by two curves: the first is the result of the mapping of the imaginary semi-axis $\omega \geq 0$ for all ω for which $j\omega$ is not a pole of $L(s)$; and the second is the symmetrical curve with respect to the real axis $\mathrm{Re}(L(j\omega))$. The first of these two curves is called the polar plot of $L(s)$, and it is a different representation of the same information contained in the Bode diagram; in fact, rather than representing magnitude and phase of the complex number $L(j\omega)$, the polar plot illustrates real and imaginary part of the complex number $L(j\omega)$.

Let us now discuss the procedure for drawing the Nyquist diagram with two simple examples.

Example 12.1

Let us consider the function $L(s) = \frac{3}{1+4s}$. For this function the Nyquist path as in Fig. 12.2(a) can be selected, as it has no poles on the imaginary axis. It does not encircle any pole in the open right half plane as $L(s)$ is stable.
The first step is to replace s with $j\omega$ and calculate the real and imaginary part of $L(j\omega)$:

$$L(j\omega) = \frac{3}{1 + j4\omega} = \frac{3(1 - j4\omega)}{(1 + j4\omega)(1 - j4\omega)} = \frac{3}{1 + 16\omega^2} - j\frac{12\omega}{1 + 16\omega^2} \qquad (12.3)$$

The real part, namely $\mathrm{Re}(L(j\omega)) = \frac{3}{1+16\omega^2}$, is always positive for any ω; it goes to zero as $\omega \to +\infty$. The imaginary part, namely $\mathrm{Im}(L(j\omega)) = -\frac{12\omega}{1+16\omega^2}$, is negative for $\omega > 0$ and equal to zero for $\omega = 0$; it goes to zero as $\omega \to +\infty$. For $\omega = 0$, the diagram starts at the point $(3, 0)$. Based on these considerations, the Nyquist diagram of $L(s)$ is the one shown in Fig. 12.3.

Example 12.2

Let us now consider $L(s) = \frac{-3}{1+4s}$. Also in this case, the function has no poles on the imaginary axis so the Nyquist path is as in Fig. 12.2(a). Also in this case, it does not encircle any pole in the open right half plane as $L(s)$ is stable.
In this case $L(j\omega)$ is given by

$$L(j\omega) = \frac{-3}{1 + j4\omega} = \frac{-3(1 - j4\omega)}{(1 + j4\omega)(1 - j4\omega)} = \frac{-3}{1 + 16\omega^2} + j\frac{12\omega}{1 + 16\omega^2} \qquad (12.4)$$

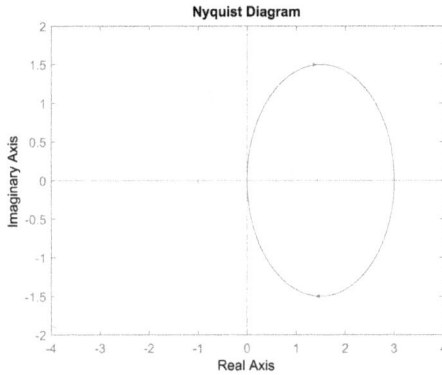

FIGURE 12.3
Nyquist diagram of $L(s) = \frac{3}{1+4s}$.

that shows that $\mathrm{Re}(L(j\omega)) = -\frac{3}{1+16\omega^2}$ and $\mathrm{Im}(L(j\omega)) = \frac{12\omega}{1+16\omega^2}$. So, the real part is always negative for any ω and goes to zero as $\omega \to +\infty$, while the imaginary part is positive for $\omega > 0$, zero for $\omega = 0$, and vanishes as $\omega \to +\infty$. For $\omega = 0$, the diagram starts at the point $(-3, 0)$. Based on these considerations, the Nyquist diagram of $L(s)$ can be drawn as in Fig. 12.4.

Let us now consider the case when $L(s)$ has one or more poles on the imaginary axis. To this aim, let us focus on the following $L(s)$:

$$L(s) = \frac{1}{s(s+1)} \tag{12.5}$$

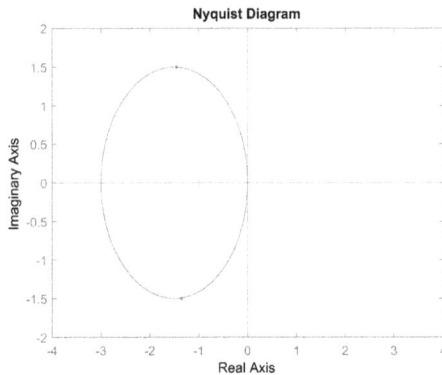

FIGURE 12.4
Nyquist diagram of $L(s) = \frac{-3}{1+4s}$.

This function has a single pole at the origin, such that the Nyquist path to consider is the one shown in Fig. 12.5.

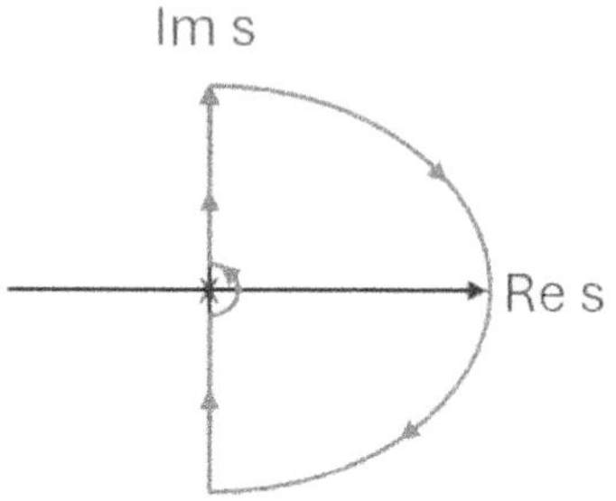

Im s

Re s

FIGURE 12.5
Nyquist path for $L(s) = \frac{1}{s(s+1)}$.

Let us now replace s with $j\omega$ and, similarly to the previous examples, calculate $L(j\omega)$ and its real and imaginary part:

$$L(j\omega) = \frac{1}{j\omega(1+j\omega)} = \frac{(1-j\omega)}{j\omega(1+j\omega)(1-j\omega)} = -\frac{1}{1+\omega^2} - j\frac{1}{\omega(1+\omega^2)} \quad (12.6)$$

The next step is to study the real and imaginary parts of $L(j\omega)$ for $\omega > 0$ and in the limit case of $\omega \to 0^+$, shortly indicated as $\omega = 0^+$. The real part, $\text{Re}(L(j\omega)) = -\frac{1}{1+\omega^2}$, is always negative for any $\omega \geq 0$. For $\omega = 0^+$, it takes the value $\text{Re}(L(j\omega)) = -1$. The imaginary part, $\text{Im}(L(j\omega)) = -\frac{1}{\omega(1+\omega^2)}$, is always negative for any $\omega > 0$. It has a discontinuity in $\omega = 0$ with $\lim_{\omega \to 0^+} \text{Im}(L(j\omega)) = -\infty$. Hence, for $\omega \to 0^+$ $L(j\omega) \to (-1, -\infty)$. The Nyquist diagram therefore displays a vertical asymptote in $\text{Re}(L(j\omega)) = -1$. For $\omega \to +\infty$ both real and imaginary parts go to zero, with a phase of $L(j\omega)$ converging to $-\pi$ as there are two poles (one at the origin and one negative) and no zeros.

For $\omega < 0$ we get a symmetric curve with $L(j\omega) \to (-1, \infty)$ for $\omega \to 0^-$ and $L(j\omega) \to (0,0)$ for $\omega \to -\infty$.

The values at the points 0^- and 0^+ can be closed by inspecting the behavior of $L(j\omega)$ in the neighborhood of $\omega = 0$. More specifically, we follow the Nyquist path by replacing s with $s = \varepsilon e^{j\theta}$ with θ taking values from $-\pi/2$ to $\pi/2$, according to the direction of the arrow in the semicircle of radius ε surrounding the origin in the Nyquist path of Fig. 12.5. By doing so, we have that $L(s) \simeq \frac{1}{\varepsilon e^{j\theta}} = \frac{1}{\varepsilon}e^{-j\theta}$. The two points 0^- and 0^+ are therefore connected by a semicircle of infinite radius $1/\varepsilon$ with phase going from $\pi/2$ to $-\pi/2$ (since the phase of $L(s)$ is $-\theta$).

The resulting Nyquist diagram is shown in Fig. 12.6.

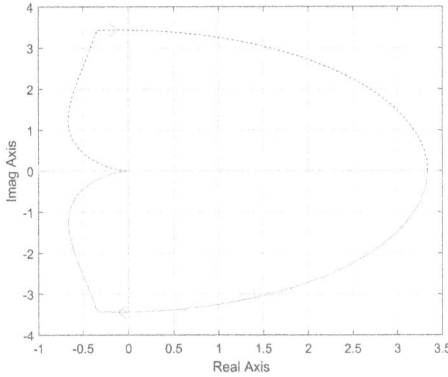

FIGURE 12.6
Nyquist diagram of $L(s) = \frac{1}{s(s+1)}$.

For imaginary poles other than the origin, the behavior is similar, except for the fact that there will be more points to inspect, as for each pair of imaginary poles in $s_{1,2} = j\pm\omega_n$, one has to consider the behavior at the points ω_n^- and ω_n^+ and the symmetrical points $-\omega_n^-$ and $-\omega_n^+$ with two semicircles of infinite radius connecting the points.

For multiple poles, e.g., in $s = 0$ with multiplicity m, we have that $L(s) \simeq \frac{1}{\varepsilon^m}e^{-jm\theta}$. Hence, the points 0^- and 0^+ will be connected by an arc of infinite radius with phase going from $m\pi/2$ to $-m\pi/2$. Several examples will be discussed after introducing the Nyquist stability criterion.

12.3 The Nyquist stability criterion

The Nyquist stability criterion offers a method to determine the stability of a closed-loop system, as in Fig. 12.1, from the characteristics of the open-loop transfer function $L(s)$, and more specifically from its Nyquist diagram.

Let P be the number of poles of $L(s)$ with positive real part, Z the number of zeros of the function $1 + L(s)$ with positive real part, i.e., the closed-loop poles in the open right half-plane, and N the net number of encirclements of the point $(-1, 0)$ by the Nyquist diagram of $L(s)$. The net number is determined by counting the counterclockwise encirclements minus the number of clockwise encirclements. These three quantities are linked by the relation

$$N = P - Z \tag{12.7}$$

such that the number of closed-loop poles in the right-half plane, namely Z, and therefore the closed-loop stability, can be determined by calculating the number of open-loop poles with positive real part, namely P, and drawing the Nyquist diagram to count the encirclements of the point $(-1, 0)$, that is, N.

From this relationship, an important necessary and sufficient condition for closed-loop stability directly follows: the closed-loop system in Fig. 12.1 is asymptotically stable if and only if N is well defined (i.e., the Nyquist diagram does not cross the critical point) and $N = P$.

Here we provide a sketch of the proof for this result. Define the function $M(s) = 1 + L(s)$. Given that $L(s) = \frac{N_L(s)}{D_L(s)}$, it follows that $M(s) = \frac{N_L(s) + D_L(s)}{D_L(s)}$. The zeros of $M(s)$ correspond to the closed-loop poles, whereas the poles of $M(s)$ coincide with the poles of $L(s)$. Therefore, $M(s)$ and $L(s)$ share the same number of poles with positive real parts, i.e., P, and the same Nyquist path.

Now, consider the Nyquist diagram of $M(s)$. Two cases can arise, which we analyze below. In the first case, while moving $s = j\omega$ along the Nyquist path for some $\bar{\omega}$ we find that the Nyquist diagram of $M(s)$ crosses the origin. Hence, $M(j\bar{\omega}) = 0$, and the closed-loop system has poles on the imaginary axis, so that it is not asymptotically stable. Note that the Nyquist diagram of $M(s)$ crossing the origin corresponds to the Nyquist diagram of $L(s)$ crossing the point $(-1, 0)$. This is the case where the number of encirclements is not well defined. In the second case, $M(s)$ does not vanish for any value of $s = j\omega$, and hence the Nyquist diagram of $M(s)$ is a closed curve that does not cross the origin. Now, let us focus on the phase of $M(s)$ while s is varied along the Nyquist path. As schematically illustrated in Fig. 12.7, any pole or zero located outside the region enclosed by the Nyquist path contributes nothing after completing the entire Nyquist path. Conversely, each pole located inside the region enclosed by the Nyquist path contributes with 2π, and each zero with -2π. Consequently, each zero inside the region causes the Nyquist diagram of $M(s)$ to make a clockwise encirclement around the origin, while each pole results in a counterclockwise encirclement. The total number of counterclockwise encirclements of the origin by $M(s)$ is therefore given by $P - Z$. Taking into account that $M(s) = 1 + L(s)$, we conclude that the total number of counterclockwise encirclements of the point $(-1, 0)$ by $M(s)$ is also given by $P - Z$. The closed-loop system is asymptotically stable if $Z = 0$, that is, when the number of zeros of $M(s)$ with positive real parts is zero. This condition requires that $N = P$.

To illustrate the application of the Nyquist stability criterion, let us consider the systems discussed in Examples 12.1 and 12.2.

Example 12.3 ⎯⎯⎯⎯⎯⎯⎯⎯⎯⎯⎯⎯⎯⎯⎯⎯⎯⎯⎯⎯⎯⎯⎯⎯⎯⎯⎯⎯⎯⎯⎯⎯

For Example 12.1, $P = 0$ since the only pole of $L(s)$ is in the open left half plane. The critical point $(-1, 0)$ is not encircled by the Nyquist diagram of $L(s)$, hence $N = 0$. It follows that $Z = 0$. The closed-loop system is asymptotically stable, as it can be

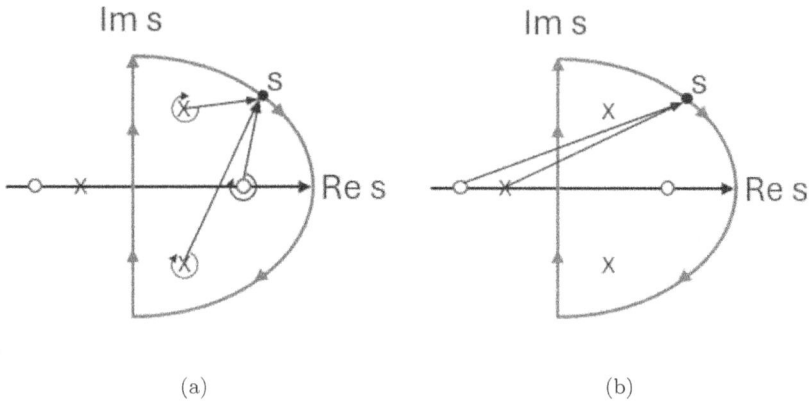

(a)   (b)

FIGURE 12.7
Schematic illustration of the contribution of poles and zeros, located inside (a) or outside (b) the region enclosed by the Nyquist path, to the phase of $M(s)$ as s traverses the entire Nyquist path.

verified by direct calculation of the closed-loop transfer function:

$$W(s) = \frac{L(s)}{1 + L(s)} = \frac{\frac{3}{1+4s}}{1 + \frac{3}{1+4s}} = \frac{3}{4 + 4s} \tag{12.8}$$

which has a negative real pole in $s = -1$.

Example 12.4 _____

For Example 12.2, the open-loop system $L(s)$ is also asymptotically stable, so $P = 0$. In this case, the critical point $(-1, 0)$ is encircled clockwise by the Nyquist diagram of $L(s)$, so $N = -1$. This yields $Z = 1$. Hence, the closed-loop system is unstable with a positive real pole. This can be verified by direct calculation of the closed-loop transfer function:

$$W(s) = \frac{L(s)}{1 + L(s)} = \frac{\frac{-3}{1+4s}}{1 - \frac{3}{1+4s}} = \frac{-3}{-2 + 4s} \tag{12.9}$$

which has a positive real pole in $s = 1/2$.

12.4   Generalization of the Nyquist stability criterion

The Nyquist stability criterion can be extended to deal with the more general case of feedback systems represented in Fig. 12.8. Here k is a real constant that,

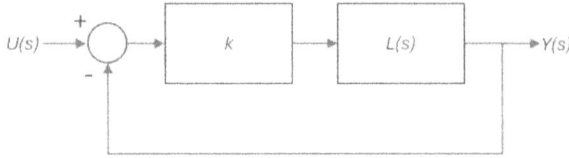

FIGURE 12.8
Feedback scheme for the study of the closed-loop stability via the generalized Nyquist stability criterion.

for instance, can represent the gain of a static controller, hence a parameter that should be tuned to guarantee stability.

The closed-loop transfer function corresponding to this diagram is given by

$$W(s) = \frac{kL(s)}{1 + kL(s)} \tag{12.10}$$

In this scenario, the closed-loop stability is therefore determined by

$$1 + kL(s) = 0 \tag{12.11}$$

This suggests to consider the point $(-1/k, 0)$ rather than $(-1, 0)$ in the analysis of the Nyquist diagram.

Hence, let P be the number of poles of $L(s)$ with positive real part, Z the number of zeros of the function $1 + L(s)$ with positive real part, i.e., the closed-loop poles in the open right half-plane, and N the net number of encirclements of the point $(-1/k, 0)$ by the Nyquist diagram of $L(s)$. Then, the generalized Nyquist stability criterion establishes that

$$N = P - Z \tag{12.12}$$

Also in this case, a direct consequence of the criterion is the following necessary and sufficient condition for closed-loop stability: the closed-loop system in Fig. 12.8 is asymptotically stable if and only if N is well defined (i.e., the Nyquist diagram does not cross the critical point) and $N = P$.

We now reexamine Examples 12.1 and 12.2 with reference to the scheme in Fig. 12.8 with $k \in \mathbb{R}$. For both systems, $P = 0$. In Example 12.1, when $k > 0$, the critical point $(-1/k, 0)$ is never encircled by the Nyquist diagram, so $N = 0$, and $Z = 0$. When $k < 0$, the critical point $(-1/k, 0)$ is not encircled if $-1/k > 3$, i.e., $k > -1/3$, while there is a clockwise encirclement of the critical point if $-1/k < 3$, i.e. $k < -1/3$. Taking altogether the analysis for positive and negative values of k, we conclude that the closed-loop system is asymptotically stable for any $k > -1/3$. This can be verified by direct calculation of the closed-loop transfer function:

$$W(s) = \frac{k\frac{3}{1+4s}}{1 + k\frac{3}{1+4s}} = \frac{3k}{3k + 1 + 4s} \tag{12.13}$$

that is indeed stable if $3k + 1 > 0$, i.e., $k > -1/3$.

In Example 12.2, we find that there is a clockwise encirclement of the critical point $(-1/k, 0)$ only if $-1/k > -3$, namely if $k > 1/3$. Thus, the closed-loop system is stable for all $k < 1/3$. In this case, the closed-loop transfer function is

$$W(s) = \frac{-k\frac{3}{1+4s}}{1 - k\frac{3}{1+4s}} = \frac{-3k}{-3k + 1 + 4s} \tag{12.14}$$

that has a negative pole if $-3k + 1 > 0$, that is, $k < 1/3$, according to the conclusion of the generalized Nyquist stability criterion.

A very important advantage of the Nyquist stability criterion is that it can be also applied to systems with a delay, e.g., $L(s) = L'(s)e^{-\tau s}$, where $\tau > 0$ is the time-delay and $L'(s)$ is rational.

12.5 Studying closed-loop stability via the Nyquist stability criterion: examples

To study the closed-loop stability via the Nyquist stability criterion, we follow these steps:

- Determine the Nyquist path.

- Compute the real and imaginary parts of $L(j\omega)$.

- Build a table providing the magnitude, phase, real and imaginary part of $L(j\omega)$ at selected values of ω. These values include $\omega = 0$ (or $\omega = 0^+$ if $s = 0$ is a pole of $L(s)$), $\omega = \infty$, and the points where the imaginary or the real part equals zero. In addition, for each $\omega_n > 0$ such that $s = j\omega_n$ is a pole of $L(s)$, the points ω_n^- and ω_n^+ are included in the table.

- Draw the Nyquist diagram.

- Apply the generalized Nyquist stability criterion for $k \in \mathbb{R}$.

Example 12.5

In this example we draw the Nyquist diagram and study the closed-loop stability for the following system

$$L(s) = \frac{10}{(s+1)(s+2)(s+3)} \qquad (12.15)$$

The system $L(s)$ has no pole on the right half plane, so $P = 0$. The Nyquist path comprises the whole imaginary axis and is as depicted in Fig. 12.2(a). We now calculate real and imaginary parts of $L(j\omega)$, as follows:

$$L(j\omega) = \frac{10}{(1+j\omega)(2+j\omega)(3+j\omega)} \qquad (12.16)$$

and so

$$
\begin{aligned}
L(j\omega) &= \frac{10(1-j\omega)(2-j\omega)(3-j\omega)}{(1+\omega^2)(4+\omega^2)(9+\omega^2)} = \\
&= \frac{60(1-\omega^2)}{(1+\omega^2)(4+\omega^2)(9+\omega^2)} + j\frac{10\omega(\omega^2-11)}{(1+\omega^2)(4+\omega^2)(9+\omega^2)}
\end{aligned} \qquad (12.17)
$$

The next step is to calculate the values of magnitude, phase, and real and imaginary parts of $L(j\omega)$ at relevant points. These always include $\omega = 0$ and $\omega = \infty$. In addition, we notice that the real part vanishes at $\omega = 1$ and the imaginary part at $\omega = \sqrt{11}$, so we include also these points in the table. The result is Table 12.1.

TABLE 12.1

Table to draw the Nyquist diagram of Example 12.5.

ω	$|L(j\omega)|$	$\angle L(j\omega)$	real $L(j\omega)$	imag $L(j\omega)$
0	$5/3$	$0°$	$5/3$	0
1	1	$-90°$	0	-1
$\sqrt{11}$	$1/6$	$-180°$	$-1/6$	0
$+\infty$	0	$-270°$	0	0

The Nyquist diagram for the system $L(s) = \frac{10}{(s+1)(s+2)(s+3)}$ is shown in Fig. 12.9. The last step is the analysis of the stability. Based on the resulting Nyquist diagram, we can distinguish four cases (recall that for this system $P = 0$):

- $-1/k < -1/6$ with $k > 0$, namely $0 < k < 6$. In this case, $N = 0$, and so $Z = 0$. The closed-loop system is asymptotically stable;

- $-1/k > -1/6$, namely $k > 6$. In this case, $N = -2$, and so $Z = 2$. The closed-loop system is unstable with two poles with positive real part;

- $-1/k > 5/3$ with $k < 0$, namely $-5/3 < k < 0$. In this case, $N = 0$, and so $Z = 0$. The closed-loop system is asymptotically stable;

- $-1/k < 5/3$, namely $k < -5/3$. In this case, $N = -1$, and so $Z = 1$. The closed-loop system is unstable with a single positive pole.

Example 12.6

Let us now consider the following system:

$$L(s) = \frac{s-2}{s(s^2-2s+10)} \qquad (12.18)$$

It has a pole at the origin, so the Nyquist path, shown in Fig. 12.10, has to exclude the pole in $s = 0$. The system also has two poles that lie in the open right half plane, also represented in Fig. 12.10. Hence $P = 2$.

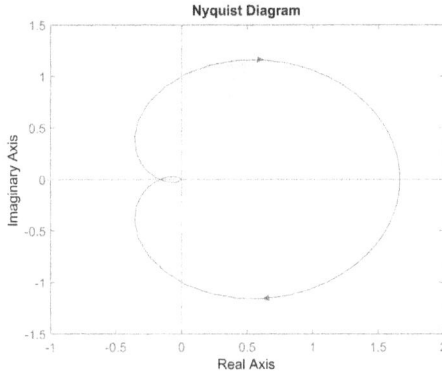

FIGURE 12.9
Nyquist diagram for the system $L(s) = \frac{10}{(s+1)(s+2)(s+3)}$.

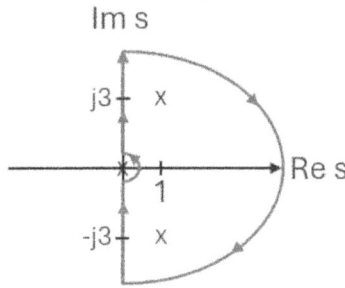

FIGURE 12.10
Nyquist path for the system $L(s) = \frac{s-2}{s(s^2-2s+10)}$.

Substituting $s = j\omega$ in $L(s)$, we get:

$$L(j\omega) = \frac{j\omega - 2}{j\omega(-\omega^2 - 2j\omega + 10)} \tag{12.19}$$

that can be rewritten as:

$$L(j\omega) = \frac{(\omega+2j)(10-\omega^2+2j\omega)}{\omega((10-\omega^2)^2+4\omega^2)} =$$
$$= \frac{6-\omega^2}{((10-\omega^2)^2+4\omega^2)} + j\frac{20}{\omega((10-\omega^2)^2+4\omega^2)} \tag{12.20}$$

The real part vanishes at $\omega = \sqrt{6}$, while the imaginary part is always positive for $\omega > 0$. Since $\omega = 0$ is a singularity for the imaginary part, Table 12.2 includes the points $\omega = 0^+$, $\omega = \sqrt{6}$, and $\omega = +\infty$.
The Nyquist diagram for the system $L(s) = \frac{s-2}{s(s^2-2s+10)}$ is shown in Fig. 12.11.

TABLE 12.2

Table to draw the Nyquist diagram of Example 12.6.

ω	$|L(j\omega)|$	$\angle L(j\omega)$	real $L(j\omega)$	imag $L(j\omega)$
0^+	$+\infty$	$90°$	$3/50$	$+\infty$
$\sqrt{6}$	$1/(2\sqrt{6})$	$90°$	0	$1/(2\sqrt{6})$
$+\infty$	0	$-180°$	0	0

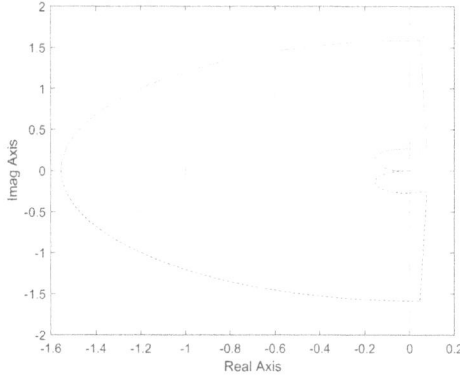

FIGURE 12.11

Nyquist diagram for the system $L(s) = \frac{s-2}{s(s^2-2s+10)}$.

The analysis of the stability is carried out taking into account that for this system $P = 2$. We have two cases:

- $k > 0$, where $N = -1$, and so $Z = 3$. The closed-loop system is unstable with three poles in the open right half plane;
- $k < 0$, where $N = 0$, and so $Z = 2$. The closed-loop system is unstable with two poles in the open right half plane.

Example 12.7 _____

In this example we consider the following system

$$L(s) = \frac{s+5}{s^2(s-1)} \tag{12.21}$$

that has two poles at the origin and a real positive one. So, $P = 1$. The Nyquist path for this system is shown in Fig. 12.12, which also includes the pole in the open right half plane.

Replacing $s = j\omega$ in $L(s)$, we get:

$$L(j\omega) = \frac{5+j\omega}{-\omega^2(j\omega-1)} \tag{12.22}$$

that can be rewritten as:

$$L(j\omega) = \frac{(5+j\omega)(1+j\omega)}{\omega^2(1+\omega^2)} = \frac{5-\omega^2}{\omega^2(1+\omega^2)} + j\frac{6\omega}{\omega^2(1+\omega^2)} \tag{12.23}$$

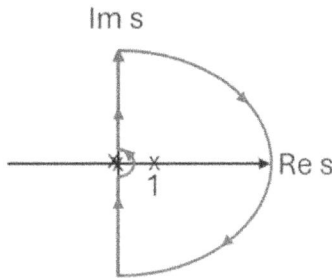

FIGURE 12.12
Nyquist path for the system $L(s) = \frac{s+5}{s^2(s-1)}$.

The real part vanishes at $\omega = \sqrt{5}$, while the imaginary part is always positive for $\omega > 0$. Hence, in Table 12.3 we consider the following points: $\omega = 0^+$, $\omega = \sqrt{5}$, and $\omega = +\infty$.

TABLE 12.3
Table to draw the Nyquist diagram of Example 12.7.

ω	$|L(j\omega)|$	$\angle L(j\omega)$	real $L(j\omega)$	imag $L(j\omega)$
0^+	$+\infty$	$0°$	$+\infty$	$+\infty$
$\sqrt{5}$	$1/\sqrt{5}$	$90°$	0	$1/\sqrt{5}$
$+\infty$	0	$-180°$	0	0

The Nyquist diagram for the system $L(s) = \frac{s+5}{s^2(s-1)}$ is shown in Fig. 12.13.

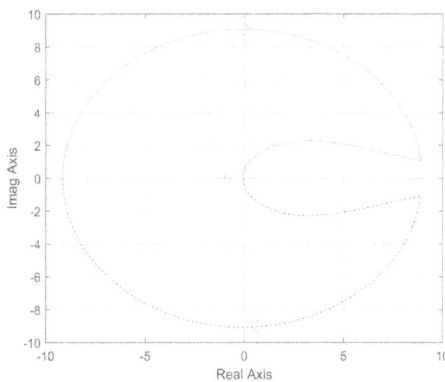

FIGURE 12.13
Nyquist diagram for the system $L(s) = \frac{s+5}{s^2(s-1)}$.

The analysis of the stability is conducted considering that, for this system, $P = 1$. We have two cases:

- $k > 0$, where $N = -1$, and so $Z = 2$. The closed-loop system is unstable with two poles in the open right half plane;

- $k < 0$, where $N = 0$, and so $Z = 1$. The closed-loop system is unstable with one pole in the open right half plane.

Example 12.8

In this example we consider the following system

$$L(s) = \frac{(1+s)^2}{s^3} \tag{12.24}$$

that has three poles at the origin, so $P = 0$. The Nyquist path for this system is shown in Fig. 12.14.

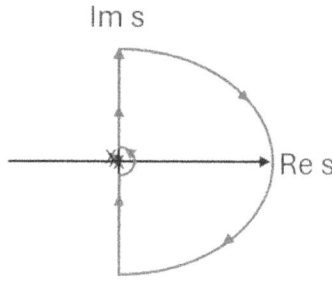

FIGURE 12.14
Nyquist path for the system $L(s) = \frac{(1+s)^2}{s^3}$.

First, we calculate and $L(j\omega)$

$$L(j\omega) = \frac{(1+j\omega)^2}{-j\omega^3} \tag{12.25}$$

and then rewrite it to calculate its real and imaginary parts:

$$L(j\omega) = \frac{1 + 2j\omega - \omega^2}{-j\omega^3} = -\frac{2}{\omega^2} + j\frac{1-\omega^2}{\omega^3} \tag{12.26}$$

We note that the real part is always negative, while the imaginary part is positive if $0 < \omega < 1$, zero if $\omega = 1$, and negative if $\omega > 1$. Relevant values of magnitude, phase, real and positive part of $L(j\omega)$ are given in Table 12.4.

TABLE 12.4
Table to draw the Nyquist diagram of Example 12.8.

ω	$|L(j\omega)|$	$\angle L(j\omega)$	real $L(j\omega)$	imag $L(j\omega)$
0^+	$+\infty$	$-270°$	$-\infty$	$+\infty$
1	2	$-180°$	-2	0
$+\infty$	0	$-90°$	0	0

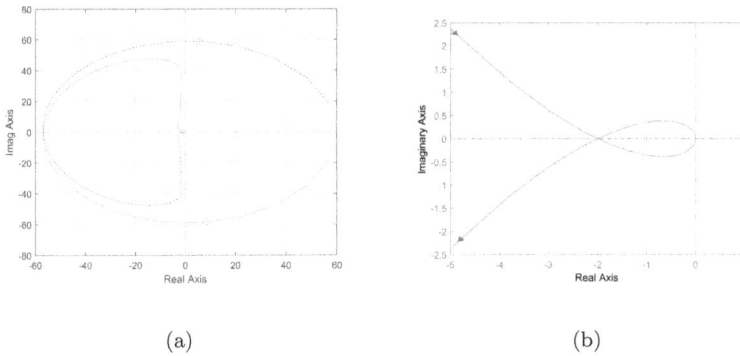

(a) (b)

FIGURE 12.15
(a) Nyquist diagram for the system $L(s) = \frac{(1+s)^2}{s^3}$. (b) Magnification of the diagram around the point $(-2,0)$.

The Nyquist diagram for the system $L(s) = \frac{(1+s)^2}{s^3}$ is shown in Fig. 12.15. The closed-loop stability is analyzed taking into account that, for this system, $P = 0$. Based on the obtained Nyquist diagram, we have three cases:

- $0 < k < 1/2$, where $N = -2$, and so $Z = 2$. The closed-loop system is unstable with two poles in the open right half plane;
- $k > 1/2$, where $N = 0$, and so $Z = 0$. The closed-loop system is asymptotically stable;
- $k < 0$, where $N = -1$, and so $Z = 1$. The closed-loop system is unstable with one pole in the open right half plane.

Example 12.9 _____

In this example we study the system

$$L(s) = \frac{s+2}{(s-2)(s^2+1)} \tag{12.27}$$

that has one positive real pole, namely $s = 2$, and a pair of poles on the imaginary axis, namely $s = \pm j$. Hence, $P = 1$. The Nyquist path for this system is shown in Fig. 12.16. Substituting $s = j\omega$ in $L(s)$, we obtain

$$L(j\omega) = \frac{2+j\omega}{(j\omega-2)(1-\omega^2)} \tag{12.28}$$

that can be rewritten as follows:

$$L(j\omega) = \frac{-(2+j\omega)(2+j\omega)}{(1-\omega^2)(2-j\omega)(2+j\omega)} = -\frac{\omega^2-4}{(1-\omega^2)(4+\omega^2)} + j\frac{-4\omega}{(1-\omega^2)(4+\omega^2)} \tag{12.29}$$

Here, we notice that the real part vanishes for $\omega = 2$. Other points to consider are 1^- and 1^+ as $L(s)$ has imaginary poles in $s_{1,2} = \pm j$. Based on these considerations, Table 12.5 gives magnitude, phase, real and positive part of $L(j\omega)$ for the following values of ω: 0, 1^-, 1^+, 2, and ∞.
The Nyquist diagram for the system $L(s) = \frac{s+2}{(s-2)(s^2+1)}$ is shown in Fig. 12.17.
For the analysis of the closed-loop stability, we consider that, for this system, $P = 1$. Based on the Nyquist diagram, we have three cases:

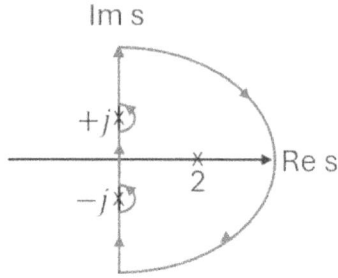

FIGURE 12.16
Nyquist path for the system $L(s) = \frac{s+2}{(s-2)(s^2+1)}$.

TABLE 12.5
Table to draw the Nyquist diagram of Example 12.9.

ω	$|L(j\omega)|$	$\angle L(j\omega)$	real $L(j\omega)$	imag $L(j\omega)$
0	1	$-180°$	-1	0
1^-	$+\infty$	$-127°$	$-\infty$	$-\infty$
1^+	$+\infty$	$-307°$	$+\infty$	$+\infty$
2	$1/3$	$-360°$	0	$1/3$
$+\infty$	0	$-180°$	0	0

- $0 < k < 1$, where $N = -2$, and so $Z = 3$. The closed-loop system is unstable with three poles in the open right half plane;

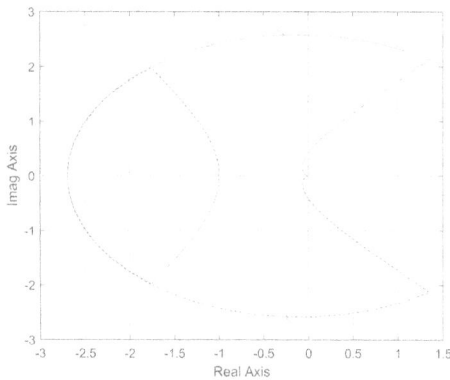

FIGURE 12.17
Nyquist diagram for the system $L(s) = \frac{s+2}{(s-2)(s^2+1)}$.

- $k > 1$, where $N = -1$, and so $Z = 2$. The closed-loop system is unstable with two poles in the open right half plane;

- $k < 0$, where $N = 0$, and so $Z = 1$. The closed-loop system is unstable with one pole in the open right half plane.

12.6 Considerations on the Nyquist stability criterion for open-loop stable systems

When the open-loop system is asymptotically stable, namely $L(s)$ has all poles in the open left half plane, we have that $P = 0$. Therefore, the closed-loop stability requires that $N = 0$, a condition that, under some circumstances, is simple to verify.

Suppose, in fact, that the open-loop system $L(s)$ is asymptotically stable, then a sufficient condition for the asymptotic stability of the closed-loop system in Fig. 12.1 is that $|L(j\omega)| < 1 \ \forall \omega$. This is a direct consequence that a Nyquist diagram featuring this property cannot encircle the point $(-1, 0)$.

In the presence of a gain k, namely for closed-loop systems as in Fig. 12.1, the sufficient condition becomes $k|L(j\omega)| < 1 \ \forall \omega$. This condition is clearly favored by a small gain k, but, as we will see later, such a small gain can have a negative impact on closed-loop performance.

Another sufficient condition for closed-loop stability is obtained by looking at the behavior of the phase rather than the magnitude. Under the hypothesis that the open-loop system $L(s)$ is asymptotically stable, a sufficient condition for the asymptotic stability of the closed-loop system in Fig. 12.1 is that $|\angle L(j\omega)| < 180° \ \forall \omega$.

12.7 Stability under perturbed conditions

So far we have considered the assumption that the open-loop model $L(s)$ is perfectly known. Suppose now that this is not the case, and the model is affected by uncertainty. From the Nyquist diagram it is possible to define two quantities that provide information on the robustness of the closed-loop stability.

The first quantity is the *gain margin*. Assume that $L(s)$ is such that its static gain μ is positive, it has no poles on the open right half plane, i.e., $P = 0$, and its Nyquist diagram crosses the x-axis only once as in Fig. 12.18(a). In this situation it follows that the closed-loop system of Fig. 12.1 is stable if

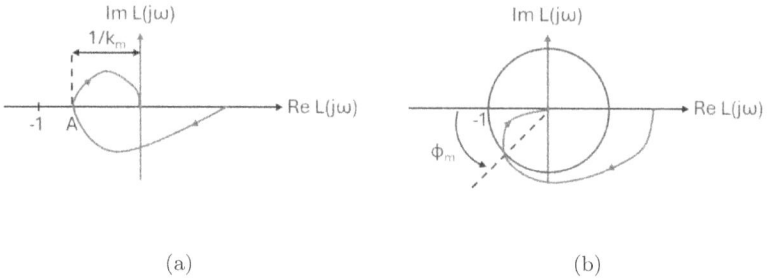

(a)   (b)

FIGURE 12.18
Stability margins: (a) gain margin k_m; (b) phase margin ϕ_m.

and only if $k_m > 1$, where k_m is the gain margin and is given by

$$k_m = \frac{1}{|L(j\omega_\pi)|} \tag{12.30}$$

with ω_π being the frequency at which $L(j\omega_\pi) = -\pi$.

Under a different perspective, if the feedback configuration also includes a tunable gain k, as in Fig. 12.8, then we can conclude that the closed-loop system is asymptotically stable for $k < k_m$, and unstable otherwise. A gain margin smaller than one, $k_m < 1$, means that the actual gain of the controller should be decreased, in order to avoid the closed-loop system is unstable, while a gain margin larger than one, $k_m > 1$, indicates that the gain of the controller can be eventually increased, e.g., to enhance the system performance. The gain margin can be expressed in dB as follows:

$$k_{m,dB} = -|L(j\omega_\pi)|_{dB} \tag{12.31}$$

with the advantage of a direct interpretation of its sign: a positive value indicates that in the actual configuration the system is stable, while a negative value that the controller gain needs to be decreased.

Lastly, the gain margin can be viewed as the uncertainty that can be tolerated on the system gain (e.g., μ) before the closed-loop system becomes unstable.

The second quantity that measures the robustness of the closed-loop stability is the *phase margin*. Suppose that $L(s)$ has a positive static gain μ, no poles on the open right half plane, i.e., $P = 0$, and its Nyquist diagram crosses the unit circle only once, from the outside to the inside, as in Fig. 12.18(b). In this scenario, the closed-loop system is asymptotically stable if and only if $\phi_m > 0$, where ϕ_m is the phase margin, given by

$$\phi_m = 180° - |\angle L(j\omega_c)| \tag{12.32}$$

where ω_c is the critical frequency at which $|L(j\omega)| = 1$.

The phase margin ϕ_m can be viewed as the uncertainty on the value of the delay τ in a system $L(s) = L'(s)e^{-\tau s}$ (with $L'(s)$ rational) that can be tolerated before the closed-loop system becomes unstable. In more detail, the phase difference introduced by the delay term is $-\omega\tau$. At most this can be equal to the phase margin, expressed in radians, i.e., $\omega_c\tau = \phi_m\pi/180°$, such that the system is stable only if $\tau < \phi_m\pi/(\omega_c 180°)$.

12.8   The Bode criterion

The Bode criterion offers a method to determine closed-loop stability for a specific class of systems without explicitly drawing the Nyquist diagram. Assume that $L(s)$ is strictly proper, has no poles with positive real parts (i.e., $P = 0$), has a positive gain ($\mu > 0$), and that its Bode diagram intersects the 0 dB axis only once. Under these conditions, a necessary and sufficient condition for stability is $m_\varphi > 0°$. These assumptions about $L(s)$ ensure that the Nyquist diagram of $L(s)$ does not encircle the point $(-1, 0)$.

12.9   Nyquist diagram in MATLAB

Nyquist diagrams can be drawn in MATLAB® using the function `nyquist`. The input of this function is an LTI object. For transfer functions $L(s)$ having poles on the imaginary axis and, thus, yielding diagrams with asymptotes, the function `lnyquist1` can be used. This function has been developed by J.N. Little and is publicly available at https://www.ece.ualberta.ca/ tchen/ct-m/extras/lnyquist1.html. It takes as inputs the numerator and denominator of the transfer function, as illustrated in the next exercise.

MATLAB® exercise 12.1 _____

Consider $L(s) = \frac{s-2}{s(s^2-2s+10)}$ as in Example 12.6. We now illustrate the commands to draw the Nyquist diagram of this system.
First, we define the transfer function

```
s=tf('s');
L=(s-2)/(s*(s^2-2*s+10));
```

Next, we use the functions `nyquist` and `lnyquist1` to draw both the numerical Nyquist diagram and the asymptotic one:

```
figure,nyquist(L)
figure,lnyquist1(L.num{:},L.den{:})
```

The asymptotic diagram obtained with `lnyquist1` is shown in Fig. 12.19.

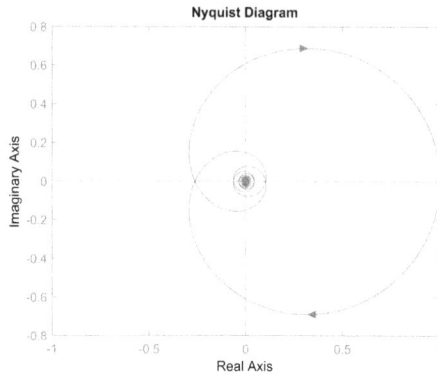

FIGURE 12.19

Nyquist diagram for the system $L(s) = \frac{1}{s+1}e^{-0.5s}$.

Another useful MATLAB® command is `margin`. This allows to obtain the gain and phase margins for a system. The function can be utilized in two ways. The first option is to generate the Bode diagrams with graphical indications of the two stability margins by using the command `margin(L)`. Alternatively, the command can be applied as follows:

```
[Gm,Pm,Wcg,Wcp] = margin(L)
```

that returns the gain margin `Gm`, the phase margin `Pm`, and the corresponding frequencies `Wcg` and `Wcp` where these margins are found.

Lastly, we illustrate an example of the Nyquist diagram of a first-order system with dead-time.

MATLAB® exercise 12.2 _____

Consider the first-order system with dead-time $L(s) = \frac{1}{s+1}e^{-0.5s}$. To draw its Nyquist diagram, the following commands can be used:

```
s=tf('s');
L=1/(s+1)*exp(-0.5*s);
w=logspace(-2,5,1001);
figure,nyquist(L,w)
```

The resulting diagram, shown in Fig. 12.19, displays the key characteristic that, due to the delay term that produces a phase continuously decreasing for $\omega \to +\infty$, the Nyquist diagram makes an infinite number of rotations around the origin. To determine the stability condition and margin, the first intersection of the diagram with the real axis must be considered.

12.10   Exercises

1. Draw the Nyquist diagram and study the closed-loop stability of the system with the following transfer function:

$$L(s) = \frac{(s+10)^2}{s(s+1)^2} \qquad (12.33)$$

2. Draw the Nyquist diagram and study the closed-loop stability of the system with the following transfer function:

$$L(s) = \frac{8(s-5)}{(s+2)^3} \qquad (12.34)$$

3. Draw the Nyquist diagram and study the closed-loop stability of the system with the following transfer function:

$$L(s) = \frac{s^2+1}{(s-2)(s+5)^2} \qquad (12.35)$$

4. Draw the Nyquist diagram and study the closed-loop stability of the system with the following transfer function:

$$L(s) = \frac{5}{(s^2+5)(1-s)} \qquad (12.36)$$

5. Draw the Nyquist diagram and study the closed-loop stability of the system with the following transfer function:

$$L(s) = \frac{10(s+2)^2}{s^2(s-5)} \qquad (12.37)$$

13

Lecture 13—Discrete-time systems: time-domain analysis

In this lecture, we introduce discrete-time systems and study their main properties in the time-domain. In doing so, we will emphasize similarities and differences with continuous-time systems. In particular, while the dynamics of continuous-time systems is defined through a set of differential equations, discrete-time systems are modeled by finite-difference equations.

13.1 State equations of a discrete-time system

In discrete-time systems, time is treated as a discrete variable, and all involved signals take values, referred to as *samples*, at specific points in time. These points are often equally spaced by a time interval known as the sampling period. This framework is particularly suitable for modeling phenomena in fields such as economics, ecology, and sociology, and for all scenarios where a digital control system is used. In the latter case, although the physical quantities are continuous in nature, they are sampled to be processed, transmitted, and managed by the digital systems that control them.

The general model to describe the state equations of a discrete-time system with m inputs, p outputs, and n state variables is the following

$$\begin{aligned} \mathbf{x}(h+1) &= f(\mathbf{x}(h), \mathbf{u}(h), h) \\ \mathbf{y}(h) &= g(\mathbf{x}(h), \mathbf{u}(h), h) \end{aligned} \tag{13.1}$$

where $\mathbf{x} \in \mathbb{R}^n$, $\mathbf{u} \in \mathbb{R}^m$, $\mathbf{y} \in \mathbb{R}^p$, $f : \mathbb{R}^n \times \mathbb{R}^m \times \mathbb{N}_0 \to \mathbb{R}^n$, $g : \mathbb{R}^n \times \mathbb{R}^m \times \mathbb{N}_0 \to \mathbb{R}^n$, and $h \in \mathbb{N}_0$. This model if formally analogous to Eqs. (2.1).

In the case of linear systems, the state equations read:

$$\begin{aligned} \mathbf{x}(h+1) &= \mathrm{A}(h)\mathbf{x}(h) + \mathrm{B}(h)\mathbf{u}(h) \\ \mathbf{y}(h) &= \mathrm{C}(h)\mathbf{x}(h) + \mathrm{D}(h)\mathbf{u}(h) \end{aligned} \tag{13.2}$$

with $\mathrm{A}(h) \in \mathbb{R}^{n \times n}$, $\mathrm{B}(h) \in \mathbb{R}^{n \times m}$, $\mathrm{C}(h) \in \mathbb{R}^{p \times n}$, and $\mathrm{D}(h) \in \mathbb{R}^{p \times m}$ for any h.

For discrete-time systems, the same terminology used for continuous-time systems is employed. Namely, we say that a system is multi-input-multi-output (MIMO), when $p > 1$ or $m > 1$, and that it is single-input-single-output

DOI: 10.1201/9781003487289-13

(SISO), when $p = 1$ and $m = 1$. Similarly, we say that Eqs. (13.1) represent a nonlinear model, while Eqs. (13.2) a linear model. Lastly, systems can be time-invariant, when f and g or the coefficients of the state matrices A, B, C, and D do not depend explicitly on time, or time-variant when the terms include an explicit dependence on time.

In the case of linear time-invariant systems, the state equations read:

$$\mathbf{x}(h+1) = \mathbf{A}\mathbf{x}(h) + \mathbf{B}\mathbf{u}(h)$$
$$\mathbf{y}(h) = \mathbf{C}\mathbf{x}(h) + \mathbf{D}\mathbf{u}(h) \tag{13.3}$$

where $\mathbf{A} \in \mathbb{R}^{n \times n}$, $\mathbf{B} \in \mathbb{R}^{n \times m}$, $\mathbf{C} \in \mathbb{R}^{p \times n}$, and $\mathbf{D} \in \mathbb{R}^{p \times m}$ are matrices of constant coefficients. This model is formally analogous to Eqs. (2.2).

Example 13.1

As an example, we consider the cohort population model due to Leslie. The state variables, $P_i(h)$ with $i = 1, \ldots, n$, represent the population at one cohort at time h. After a time unit, a fraction a_i of a cohort moves to the next cohort, that is, $P_{i+1}(h + 1) = a_i P_i(h)$ for $i = 1, \ldots, n-1$. For the first cohort we have that $P_1(h+1) = b_1 P_1(h) + b_2 P_2(h) + \ldots + b_n P_n(h)$, where the coefficients b_i are the birth rates of the different cohorts. The model includes an input for each cohort, which represents individuals joining the cohort from outside. The output of the model is the total population $y_h = P_1(h) + \ldots + P_n(h)$.

This model is in the form of Eqs. (13.3) with the following state matrices:

$$A = \begin{bmatrix} b_1 & b_2 & \cdots & \cdots & b_n \\ a_1 & 0 & \cdots & \cdots & 0 \\ 0 & a_2 & 0 & \cdots & 0 \\ & & \ddots & & \\ 0 & 0 & & 0 & 0 \\ 0 & 0 & \cdots & a_{n-1} & 0 \end{bmatrix} ; \quad B = I$$
$$C = \begin{bmatrix} 1 & 1 & \cdots & \cdots & 1 \end{bmatrix} ; \quad D = 0 \tag{13.4}$$

Eqs. (13.3) represent a state-space representation of a discrete-time system. Similarly to continuous-time systems, this representation is not unique and different state variables may be used. Considering a new set of state variables $\tilde{\mathbf{x}}$ such that $\tilde{\mathbf{x}} = T^{-1}\mathbf{x}$, we can obtain an equivalent representation in state-space form:

$$\tilde{\mathbf{x}}(h+1) = \tilde{\mathbf{A}}\tilde{\mathbf{x}}(h) + \tilde{\mathbf{B}}\mathbf{u}(h)$$
$$\mathbf{y}(h) = \tilde{\mathbf{C}}\tilde{\mathbf{x}}(h) + \tilde{\mathbf{D}}\mathbf{u} \tag{13.5}$$

where

$$\tilde{\mathbf{A}} = T^{-1}\mathbf{A}T$$
$$\tilde{\mathbf{B}} = T^{-1}\mathbf{B}$$
$$\tilde{\mathbf{C}} = \mathbf{C}T \tag{13.6}$$
$$\tilde{\mathbf{D}} = \mathbf{D}$$

Also for discrete-time systems A and $\tilde{\mathbf{A}}$ are linked by a similarity relationship, i.e., $\tilde{\mathbf{A}} = T^{-1}\mathbf{A}T$, and then have the same eigenvalues that are called the *system eigenvalues*.

13.2 Equilibrium points

The concept of an equilibrium point is the same for continuous-time and discrete-time systems. In both cases, equilibrium points represent constant solutions to the system equations. However, in one case, a constant solution implies that the derivative of the state variables is zero, while, in the other, it means that the state variables assume the same value at each iteration. Therefore, for discrete-time systems, we have to consider the condition $\mathbf{x}(h + 1) = \mathbf{x}(h) = \bar{\mathbf{x}}$ in the presence of a constant input $\bar{\mathbf{u}}$, which, replaced in the first of Eqs. (13.1), leads to

$$\bar{\mathbf{x}} = f(\bar{\mathbf{x}}, \bar{\mathbf{u}}) \tag{13.7}$$

For discrete-time LTI systems (13.3), this condition becomes

$$\bar{\mathbf{x}} = A\bar{\mathbf{x}} + B\bar{\mathbf{u}} \tag{13.8}$$

If $I - A$ is invertible, namely if A does not have any eigenvalue equal to 1, then, in correspondence with the input $\bar{\mathbf{u}}$, there exists a unique equilibrium point $\bar{\mathbf{x}}$ given by

$$\bar{\mathbf{x}} = (I - A)^{-1} B\bar{\mathbf{u}} \tag{13.9}$$

Otherwise, there are two possible scenarios: no solution exists or Eq. (13.8) has infinite solutions.

13.3 Time evolution of the state variables and of the output

For discrete-time LTI systems, the time evolution of the state variables and the output can be calculated with two expressions analogous to the ones that we have discussed for continuous-time LTI systems in Sec. 5.2.

Let us consider a discrete-time LTI system (13.3) that, at time $h_0 = 0$, is in the state $\mathbf{x}(0) = \mathbf{x}_0$, then the time evolution of the state variables, namely $\mathbf{x}(h)$ for $h = 1, 2, \ldots$, can be expressed through the following formula:

$$\mathbf{x}(h) = A^h \mathbf{x}(0) + \sum_{i=0}^{h-1} A^{h-i-1} B\mathbf{u}(i) \tag{13.10}$$

and similarly the output $\mathbf{y}(h)$ is given by

$$\mathbf{y}(h) = CA^h \mathbf{x}(0) + C \sum_{i=0}^{h-1} A^{h-i-1} B\mathbf{u}(i) + D\mathbf{u}(h) \tag{13.11}$$

Comparing these expressions with the Lagrange's formulas (5.13) and (5.14), we note that the matrix A^h plays the role of the exponential matrix e^{At}.

The relationship (13.10) can be proved by iterating the calculation of $\mathbf{x}(h)$ from (13.3). In fact, letting $h = 0$ in (13.3) we derive that

$$\mathbf{x}(1) = A\mathbf{x}(0) + B\mathbf{u}(0) \tag{13.12}$$

Next, considering $h = 1$ we find

$$\mathbf{x}(2) = A\mathbf{x}(1) + B\mathbf{u}(1) \tag{13.13}$$

that, using (13.12), becomes

$$\mathbf{x}(2) = A^2\mathbf{x}(0) + AB\mathbf{u}(0) + B\mathbf{u}(1) \tag{13.14}$$

and so on to calculate $\mathbf{x}(3)$, $\mathbf{x}(4)$, ...

Similar considerations hold for Eq. (13.11).

Eqs. (13.10) and (13.11) are of great theoretical importance. However, to compute $\mathbf{x}(h)$ and $\mathbf{y}(h)$ it is more convenient to proceed as we did in (13.12) and (13.13) and so on, using iteratively Eqs. (13.3).

Similarly to what observed for the Lagrange's formula, also Eqs. (13.10) and (13.11) can be rewritten as the sum of two terms:

$$\mathbf{x}(h) = \mathbf{x}_n(h) + \mathbf{x}_f(h) \tag{13.15}$$

where the first term represents the natural evolution of the state

$$\mathbf{x}_n(h) = A^h\mathbf{x}(0) \tag{13.16}$$

and the second one the forced evolution

$$\mathbf{x}_f(h) = \sum_{i=0}^{h-1} A^{h-i-1}B\mathbf{u}(i) \tag{13.17}$$

Analogously, for the output we have that

$$\mathbf{y}(h) = \mathbf{y}_n(h) + \mathbf{y}_f(h) \tag{13.18}$$

where the first term is the natural response

$$\mathbf{y}_n(h) = CA^h\mathbf{x}(0) \tag{13.19}$$

and the second is the forced response

$$\mathbf{y}_f(h) = C\sum_{i=0}^{h-1} A^{h-i-1}B\mathbf{u}(i) + D\mathbf{u}(h) \tag{13.20}$$

13.4   Superposition principle

The superposition principle also applies to discrete-time linear systems. While discussed here in the context of LTI systems, this property is more general and holds for any linear system.

Let $\mathbf{x}'(h)$ and $\mathbf{y}'(h)$ be the state and output evolution of system (13.3) starting from initial conditions $\mathbf{x}'_0$ and subject to input $\mathbf{u}'(h)$, and let $\mathbf{x}''(h)$ and $\mathbf{y}''(h)$ be the state and output evolution of system (13.3) starting from initial conditions $\mathbf{x}''_0$ and subject to input $\mathbf{u}''(h)$, then, for any pair of values of α and β, the time evolution of the state and the output of the system starting from an initial condition given by $\alpha \mathbf{x}'_0 + \beta \mathbf{x}''_0$ and subject to the input $\alpha \mathbf{u}'(h) + \beta \mathbf{u}''(h)$ are given by

$$\mathbf{x}'''(h) = \alpha \mathbf{x}'(h) + \beta \mathbf{x}''(h) \tag{13.21}$$

and

$$\mathbf{y}'''(h) = \alpha \mathbf{y}'(h) + \beta \mathbf{y}''(h) \tag{13.22}$$

that is, they can be obtained by evaluating separately the effect of the initial condition $\mathbf{x}'_0$ and the input $\mathbf{u}'(h)$ and the effect of the initial condition $\mathbf{x}''_0$ and the input $\mathbf{u}''(h)$, and then considering the sum of the two effects, weighted by α and β.

13.5   Stability and system eigenvalues

The notion of stability does not depend on the nature of the system, such that for discrete-time systems the same definitions and concepts hold. In particular, the equilibrium points of a nonlinear discrete-time system can be asymptotically stable, asymptotically stable, or unstable. For discrete-time LTI systems all equilibrium points have the same characteristics of stability, such that stability becomes a property of the system itself. Also in this case, the system stability depends on the behavior of the natural evolution, such that a discrete-time LTI system is asymptotically stable if and only if:

$$\lim_{h \to +\infty} A^h = 0 \tag{13.23}$$

The stability of a discrete-time LTI system thus depends on the eigenvalues of the state matrix A, that is, on the system eigenvalues. The remarkable difference between continuous-time and discrete-time systems is that for continuous-time systems we have to consider the exponential matrix e^{At}, while for discrete-time systems the power matrix A^h. This yields different criteria on the system eigenvalues in the two cases.

For discrete-time LTI systems, indeed, the modes are of the type: λ_i^h, $h\lambda_i^{h-1}$, $h^2\lambda_i^{h-2}$, ..., if λ_i is real, and $\rho_i^h \sin(\theta_i h)$, $\rho_i^h \cos(\theta_i h)$, $h\rho_i^{h-1} \sin(\theta_i(h-1))$, $h\rho_i^{h-1} \cos(\theta_i(h-1))$ ..., if $\lambda_i = \rho_i e^{j\theta_i}$ is complex. Hence, if and only if all eigenvalues have a modulus less than one, all modes go to zero for $h \to +\infty$, and so does A^h, and the system is asymptotically stable.

13.6   Time evolution of the system modes

It is interesting to qualitatively analyze the time evolution of the discrete-time system's modes based on the location of its eigenvalues. Let us first consider the case $n = 1$, i.e., a scalar discrete-time system. The system has a single eigenvalue, which can be located inside, outside, or on the unit circle. As discussed earlier, the stability of the system depends on the modulus of the eigenvalue: when the modulus is less than one, i.e., the eigenvalue lies within the unit circle, the equilibrium is asymptotically stable. However, the sign of the eigenvalue plays a crucial role in how the system converges to the equilibrium. Let us consider two cases differing for the value of A, namely $A = 0.8$, and $A = -0.8$. The time evolutions of the modes starting from $x(0) = 1$ are illustrated in the upper panel of Fig. 13.1. For positive eigenvalues, the mode

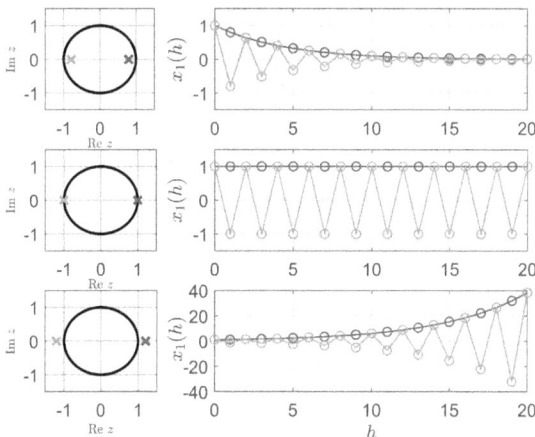

FIGURE 13.1
Time evolution of the modes of scalar LTI discrete-time systems based on the location of the system eigenvalue.

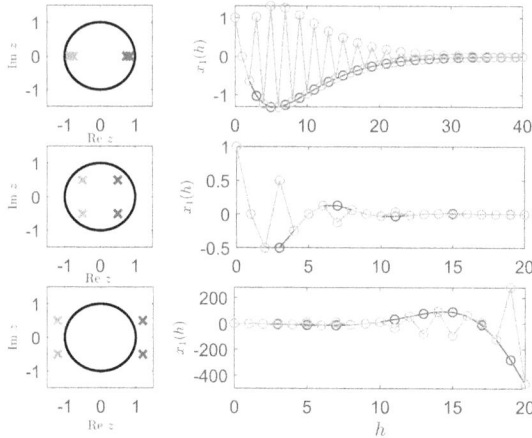

FIGURE 13.2
Time evolution of the modes of LTI discrete-time systems having eigenvalues
with double multiplicity or a pair complex and conjugate eigenvalues with
modulus smaller or larger than one.

converges monotonically to the equilibrium, whereas for negative eigenvalues,
the mode oscillates around the equilibrium with decreasing amplitude. The
behavior of the modes for $A = 1$ or $A = -1$ is instead illustrated in the central
panel of Fig. 13.1: while for $A = 1$ the mode remains constant at its initial
value $x(0) = 1$, for $A = -1$ the system oscillates. Finally, instability occurs for
system eigenvalues located outside the unit circle, with monotonically diverg-
ing modes for positive eigenvalues and oscillations for negative eigenvalues,
as shown in the lower panel of Fig. 13.1 that corresponds to $A = 1.2$ and
$A = -1.2$.

When $n > 1$, the system can have multiple eigenvalues that give rise to
the time evolutions illustrated in Fig. 13.2 for $\mathbf{x}(0) = \begin{bmatrix} 1 \\ 0 \end{bmatrix}$. For instance, a
real eigenvalue located inside the unit circle and having double multiplicity is
associated with asymptotically stable modes that are non-monotonic for positive eigenvalues, and oscillatory with non-monotonic amplitudes for negative
eigenvalues, as shown in the upper panel of Fig. 13.2.

Damped oscillations are found also in the case of complex and conjugate
eigenvalues with a modulus smaller than one, as shown in the central panel
of Fig. 13.2, while if the modulus is larger than one the oscillations display
increasing amplitude, as shown in the lower panel of Fig. 13.2.

When the system has two eigenvalues on the unit circle, the modes can
display the behavior illustrated in Fig. 13.3, where the upper panel represents
the case of real eigenvalues with double multiplicity, the central panel the case

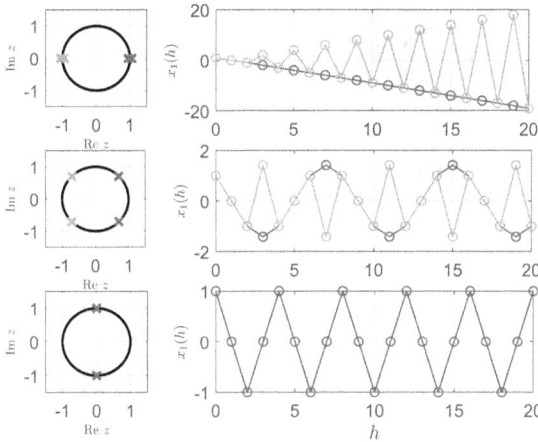

FIGURE 13.3
Time evolution of the modes of LTI discrete-time systems having two eigenvalues located on the unit circle.

of complex and conjugate eigenvalues with unit modulus, and the lower panel the case of a pair of purely imaginary eigenvalues with unit modulus, namely $\lambda_{1,2} = \pm j$.

A distinctive behavior arises when all the system eigenvalues are zero. In this scenario, as illustrated in Fig. 13.4, the modes reach the equilibrium in a finite amount of time, determined by the eigenvalue multiplicity. This phenomenon is unique to discrete-time systems and does not occur in continuous-time dynamics, where stable modes approach the equilibrium asymptotically rather than in finite time. Systems exhibiting this behavior are known as Finite Impulse Response (FIR) systems and will be discussed in more detail in the next lecture.

13.7 Reachability, controllability and observability in discrete-time systems

The notions of reachable, controllable, and observable, states that have been discussed in Lecture 9 for continuous-time systems also apply to discrete-time systems. There is, however, an important difference. While continuous-time systems are always reversible, that is, a trajectory produced by such systems remains a solution of the system when time is reversed, the same is not true for all discrete-time systems. The reversibility of a system is associated with

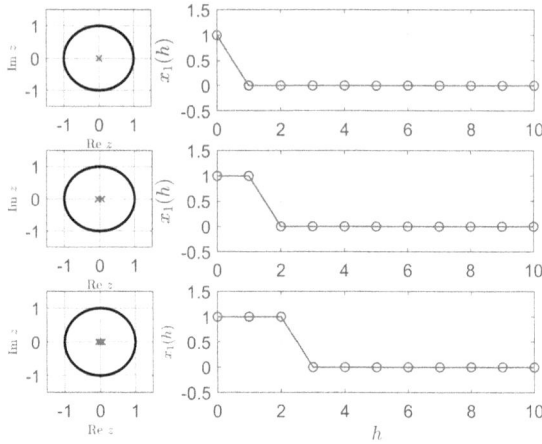

FIGURE 13.4
Time evolution of the modes of LTI discrete-time systems having eigenvalues at the origin.

the invertibility of the matrix e^{At} in the continuous-time case and A^h in the discrete-time case. For example, a FIR system, having all eigenvalues of A at the origin, is not reversible. Since all natural evolutions converge to zero in a finite number of steps, the system is completely controllable. However, there is no guarantee that it is also completely reachable. While for continuous-time systems reachability implies controllability and viceversa, for discrete-time systems reachability implies controllability but not viceversa.

Example 13.2 _____

Consider the following discrete-time LTI system:

$$A = \begin{bmatrix} 0 & 0 \\ 0 & 0 \end{bmatrix}; \quad B = \begin{bmatrix} 1 \\ 0 \end{bmatrix};$$
$$C = \begin{bmatrix} 1 & 1 \end{bmatrix}; \quad D = 0 \tag{13.24}$$

This is a FIR system. It is completely controllable as the natural evolution from any initial condition reaches zero after one step. However, the system is not completely reachable, as the reachability matrix $M_r = \begin{bmatrix} 1 & 0 \\ 0 & 0 \end{bmatrix}$ is not full rank.

13.8   Sampled-data systems

Control systems are often implemented in digital technology. Controlling a continuous-time system with a discrete-time controller requires deriving an equivalent discrete-time representation of the system to be controlled. Such systems are referred to as sampled-data systems.

Let us consider a continuous-time system represented in the time-domain as in Eqs. (2.2), with state matrices A, B, C, and D, and a digital control system with sampling time T. Since the control action remains constant between two successive samples of the digital control system, i.e., between hT and $(h+1)T$, a discrete-time realization of the continuous-time system can be obtained by integrating Eqs. (2.2) over the interval $[hT, (h+1)T]$ as follows:

$$\mathbf{x}((h+1)T) = e^{AT}\mathbf{x}(hT) + \int_{hT}^{(h+1)T} e^{A((h+1)T-\tau)}\mathbf{Bu}(\tau)d\tau \qquad (13.25)$$

yielding the following discrete-time representation

$$\begin{aligned}\mathbf{x}(h+1) &= A_s\mathbf{x}(h) + B_s\mathbf{u}(h) \\ \mathbf{y}(h) &= \mathbf{Cx}(h) + \mathbf{Du}(h)\end{aligned} \qquad (13.26)$$

with

$$\begin{aligned}A_s &= e^{AT} \\ B_s &= \int_0^T e^{A\tau}d\tau B\end{aligned} \qquad (13.27)$$

If $\det A \neq 0$, then $B_s = A^{-1}\left(e^{AT} - I\right)B$. The system in Eqs. (13.26) is called sampled-data system.

13.9   Exercises

1. Study the stability of the discrete-time LTI system with the following state matrix A

$$A = \begin{bmatrix} 0.3 & -0.4 \\ -0.1 & 0.3 \end{bmatrix} \qquad (13.28)$$

2. Study the stability of the discrete-time LTI system with the following state matrix A

$$A = \begin{bmatrix} 0.2 & 0.5 \\ -1 & -0.2 \end{bmatrix} \qquad (13.29)$$

3. Determine for which values of $k \in \mathbb{R}$ the discrete-time LTI system with the following state matrix A

$$A = \begin{bmatrix} 0 & 1 \\ -0.2k & k+0.2 \end{bmatrix} \qquad (13.30)$$

is stable.

4. Calculate and plot in MATLAB® the modes of a discrete-time LTI system with the following state matrix A

$$A = \begin{bmatrix} -1 & 0 & 0 \\ 0 & -2 & 0 \\ 0 & 0 & 2 \end{bmatrix} \tag{13.31}$$

5. Consider the following continuous-time LTI system

$$A = \begin{bmatrix} 0 & 1 \\ -1 & -1 \end{bmatrix}; \quad B = \begin{bmatrix} 0 \\ 1 \end{bmatrix}; \tag{13.32}$$
$$C = \begin{bmatrix} 1 & 2 \end{bmatrix}; \quad D = 0$$

Calculate a sampled-data system, using a sampling time fixed at $T = 0.1$ s.

14

Lecture 14—Discrete-time systems: z-transform and the design of discrete-time control systems

In this lecture, we revisit the discussion on discrete-time systems within the z-transform domain. The fundamental concepts of the z-transform are introduced, including the notion of discrete-time transfer function. The relationship between the Laplace and z-transforms is also explored, providing a framework for mapping continuous-time systems to their discrete-time counterparts. Finally, strategies for designing discrete-time control systems are presented.

14.1   The z-transform

In the previous lectures, we introduced the concept of the Laplace transform to address continuous-time LTI systems. This transformation maps the problem of solving a differential equation into an algebraic one, making it more tractable and easier to solve. In this lecture, we focus on defining a different transformation that similarly converts the finite-difference equations describing a discrete-time system into an algebraic problem.

The z-transform of a discrete-time signal $x(h)$ with $h = 0, 1, \ldots$ is defined as

$$X(z) = \sum_{h=0}^{\infty} x(h) z^{-h} \tag{14.1}$$

where z is a complex variable, often expressed in polar form as $z = \rho e^{j\phi}$.

Similarly to what was done for the Laplace transform (see Lecture 4), we will restrict the discussion on a few properties that are relevant to the analysis and control of linear systems, without any pretense of completeness.

For this reason, although rigorously Eq. (14.1) requires the convergence of the series, which leads to the definition of the radius of convergence, we will assume to deal with signals such that the z-transform can be defined (or, more precisely, its extension) for any z.

DOI: 10.1201/9781003487289-14

The z-transform has many properties similar to those of the Laplace transform and, for many aspects, it plays a complementary role for discrete-time signals with respect to the one that the Laplace transform has for continuous-time signals.

- The z-transform is a linear operator. Given two real numbers a_1 and a_2, if $x(h) = a_1 x_1(h) + a_2 x_2(h)$, then

$$X(z) = a_1 X_1(z) + a_2 X_2(z) \qquad (14.2)$$

- The z-transform of a time-delayed signal, that is, $x(h) = x_1(h-1)$, is given by

$$X(z) = z^{-1} X_1(z) \qquad (14.3)$$

More in general if $x(h) = x_1(h-h_1)$, where h_1 is an integer and $x_1(h-h_1) = 0$ for $h < h_1$, then

$$X(z) = z^{-h_1} X_1(z) \qquad (14.4)$$

- The z-transform of $x(h) = x_1(h)a^h$ is

$$X(z) = X_1\left(\frac{z}{a}\right) \qquad (14.5)$$

- The z-transform of $x(h) = x_1(h+1)$ is

$$X(z) = z X_1(z) - z x_1(0) \qquad (14.6)$$

More in general if $x(h) = x_1(h+n)$, where n is an integer, then

$$X(z) = z^n X_1(z) - z^n x_1(0) - z^{n-1} x_1(1) - \ldots - z x_1(n-1) \qquad (14.7)$$

Table 14.1 illustrates the z-transform of several relevant signals.

14.2   Relationship between the z-domain and the Laplace domain

It is useful to explore the relationship between the z-transform domain and the Laplace transform domain, as it offers valuable insights into how the properties of LTI continuous-time systems translate into the digital domain.

The z-transform of a signal, as in Eq. (14.1), can be obtained by sampling the Laplace transform (4.1) of the signal, i.e., replacing $z = e^{sT}$. Given $s = \sigma + j\omega$, we have that

TABLE 14.1

List of common z-transforms.

$x(h)$	$X(z)$
1	$\frac{z}{z-1}$
a^h	$\frac{z}{z-a}$
ha^{h-1}	$\frac{z}{(z-a)^2}$
h	$\frac{z}{(z-1)^2}$
$\begin{cases} a^{h-1} & h \geq 1 \\ 0 & h = 0 \end{cases}$	$\frac{1}{z-a}$
$\begin{cases} 1 & h = 0 \\ 0 & h > 0 \end{cases}$	1

$$z = e^{\sigma T}e^{j\omega T} \tag{14.8}$$

Therefore, it is possible to evaluate the modulus and the phase of z as

$$\begin{aligned} |z| &= e^{\sigma T} \\ \angle z &= \omega T \end{aligned} \tag{14.9}$$

The relationships (14.9) allow to determine the correspondence between the z-plane and the s-plane, such that

$$\begin{aligned} \sigma < 0 & \quad |z| < 1 \\ \sigma = 0 & \quad |z| = 1 \\ \sigma > 0 & \quad |z| > 1 \end{aligned} \tag{14.10}$$

The condition for asymptotic stability in the continuous-time domain, where $\sigma < 0$, maps to the interior of the unit circle in the discrete-time domain, i.e., $|z| < 1$. Similarly, the imaginary axis in the s-plane, where $\sigma = 0$, corresponds to the unit circle itself, i.e., $|z| = 1$, as schematically illustrated in Fig. 14.1.

14.3 Transfer function

Consider the state-space representation for a discrete-time LTI system, provided here for convenience:

$$\begin{aligned} \mathbf{x}(h+1) &= \mathbf{A}\mathbf{x}(h) + \mathbf{B}\mathbf{u}(h) \\ \mathbf{y}(h) &= \mathbf{C}\mathbf{x}(h) + \mathbf{D}\mathbf{u}(h) \end{aligned} \tag{14.11}$$

and assume $\mathbf{x}(0) = 0$. Applying the z-transform to both sides of Eqs. (14.11), we obtain

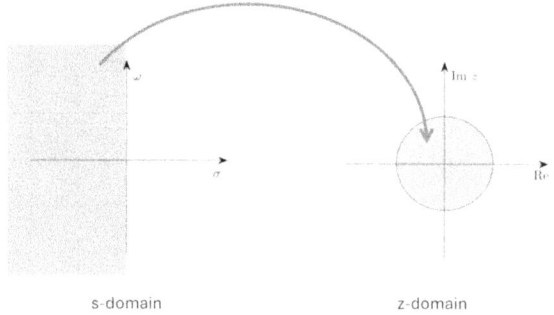

s-domain   z-domain

FIGURE 14.1
Mapping of the s-plane to the z-plane, illustrating the correspondence between continuous-time and discrete-time stability regions.

$$
\begin{aligned}
z\mathbf{X}(z) &= \mathbf{A}\mathbf{X}(z) + \mathbf{B}\mathbf{U}(z) \\
\mathbf{Y}(z) &= \mathbf{C}\mathbf{X}(z) + \mathbf{D}\mathbf{U}(z)
\end{aligned}
\tag{14.12}
$$

From these equations, we get

$$
\mathbf{Y}(z) = \left[\mathbf{C}\left(z\mathbf{I} - \mathbf{A}\right)^{-1}\mathbf{B} + \mathbf{D} \right]\mathbf{U}(z)
\tag{14.13}
$$

Therefore, the transfer matrix/function of a discrete-time LTI system is given by

$$
\mathbf{G}(z) = \mathbf{C}\left(z\mathbf{I} - \mathbf{A}\right)^{-1}\mathbf{B} + \mathbf{D}
\tag{14.14}
$$

The expression is analogous to the one that we have found for continuous-time LTI systems. Other properties apply as well. In fact, given a transformation of the state variables of the type $\tilde{\mathbf{x}} = \mathbf{T}^{-1}\mathbf{x}$, then, in virtue of Eqs. (13.5), we find that $\tilde{\mathbf{G}}(z) = \mathbf{G}(z)$. This allows us to conclude that the transfer matrix is a system invariant.

In analogy to what is found for continuous-time LTI systems, also for discrete-time LTI systems, it is immediate to derive that the transfer function is the z-transform of the impulse response.

Lastly, we note that, given an input $u(k)$, we can calculate the corresponding system response, by using $Y(z) = G(z)U(z)$ and then applying the inverse z-transform. In doing so, we use an approach similar to that adopted for continuous-time LTI systems. The difference is that, in the discrete-time case, it is more convenient to expand in simple fractions $Y(z)/z$, rather than $Y(z)$, as many common z-transforms (see, for instance, Table 14.1) have a numerator equal to z. The next example illustrates this point.

Example 14.1

Let us consider the discrete-time LTI system described by the following equation:

$$x(h+2) = x(h+1) + x(h) \tag{14.15}$$

with $x(0) = 0$ and $x(1) = 1$, and let us find the general expression of the state variable $x(h)$ as a function of time h. To this aim, we first apply the z-transform to both sides of Eq. (14.15). We obtain

$$z^2 X(z) - z^2 x(0) - zx(1) = zX(z) - zx(0) + X(z) \tag{14.16}$$

Replacing the values $x(0) = 0$ and $x(1) = 1$ and solving for $X(z)$ we get

$$X(z) = \frac{z}{z^2 - z - 1} \tag{14.17}$$

To calculate the inverse z-transform, we consider $\frac{X(z)}{z}$ and expand it in simple fractions as follows:

$$\frac{X(z)}{z} = \frac{1}{z^2 - z - 1} = \frac{r_1}{z - \frac{1+\sqrt{5}}{2}} + \frac{r_2}{z - \frac{1-\sqrt{5}}{2}} \tag{14.18}$$

The two residues r_1 and r_2 are given by

$$r_1 = \lim_{z \to \frac{1+\sqrt{5}}{2}} \left(z - \frac{1+\sqrt{5}}{2} \right) \frac{1}{z^2 - z - 1} = \lim_{z \to \frac{1+\sqrt{5}}{2}} \frac{1}{z - \frac{1-\sqrt{5}}{2}} = \frac{1}{\sqrt{5}} \tag{14.19}$$

and

$$r_2 = \lim_{z \to \frac{1-\sqrt{5}}{2}} \left(z - \frac{1-\sqrt{5}}{2} \right) \frac{1}{z^2 - z - 1} = \lim_{z \to \frac{1+\sqrt{5}}{2}} \frac{1}{z - \frac{1+\sqrt{5}}{2}} = -\frac{1}{\sqrt{5}} \tag{14.20}$$

Substituting these values into (14.18) we obtain

$$\frac{X(z)}{z} = \frac{1}{\sqrt{5}} \left(\frac{1}{z - \frac{1+\sqrt{5}}{2}} - \frac{1}{z - \frac{1-\sqrt{5}}{2}} \right) \tag{14.21}$$

and so

$$X(z) = \frac{1}{\sqrt{5}} \left(\frac{z}{z - \frac{1+\sqrt{5}}{2}} - \frac{z}{z - \frac{1-\sqrt{5}}{2}} \right) \tag{14.22}$$

Using the results of Table 14.1, we derive that

$$x(h) = \frac{1}{\sqrt{5}} \left[\left(\frac{1+\sqrt{5}}{2} \right)^h - \left(\frac{1-\sqrt{5}}{2} \right)^h \right] \tag{14.23}$$

Notice that this expression does not clearly show that all values of $x(h)$ are integers for any $h = 0, 1, 2, \ldots$, which is instead evident from Eq. (14.15) where each sample is computed as the sum of two previous samples. In fact, if one is interested in the whole evolution of $x(h)$ it is more convenient to apply Eq. (14.15) iteratively, rather than Eq. (14.23). The sequence of numbers generated this way corresponds to the well-known series of Fibonacci numbers.

14.4   Final value theorem in the z-domain

We have seen in Lecture 7 that the final value of a signal in the time-domain can be determined from its Laplace transform by means of the final value theorem. A similar property holds in the z-domain. Given the discrete-time signal $y(hT)$, and assuming that the limit for $h \to \infty$ exists and it is finite, it holds that

$$\lim_{h \to +\infty} y(hT) = \lim_{z \to 1} (z - 1) Y(z) \tag{14.24}$$

Let us consider a BIBO stable discrete-time system with transfer function $G(z)$. In order to determine the final value of the unit step response, that is the static gain of the transfer function, we can apply the final value theorem. Recalling that the z-transform of the unit step is $U(z) = \frac{z}{z-1}$, the step response is $Y(z) = \frac{z}{z-1} G(z)$. The final value theorem yields $Y_\infty = \lim_{z \to 1} (z - 1) Y(z) = \lim_{z \to 1} z G(z) = G(1)$.

14.5   The bilinear transform

The bilinear transform $z = \frac{1+s}{1-s}$ is used to transform a discrete-time system in a continuous-time system, and viceversa via the inverse transform $s = \frac{z-1}{z+1}$. The bilinear transform is particularly useful to study the stability of a system from the characteristic polynomial of matrix A or the denominator of the transfer function, since it maps points that lie inside the unit circle into points of the complex left half-plane. Hence, applying this transform we can study the stability of a discrete-time system via the analysis of an equivalent continuous-time system. Once the equivalent continuous-time system is obtained, the criteria and methods holding for continuous-time systems may be used. For instance, let $p(z)$ be the denominator of the transfer function $G(z)$ of a discrete-time LTI system. Then, replacing $z = \frac{1+s}{1-s}$ into $p(z)$, we obtain $p(\frac{1+s}{1-s})$. We can study the position of the roots of this polynomial with the criteria that apply to continuous-time systems, e.g., the Routh criterion, to understand whether the original (discrete-time) system is stable or not.

Example 14.2 _____

Let us consider a discrete-time system with a transfer function whose denominator is $p(z) = z^2 - z - 1$. Such a system is not stable as the roots of the polynomial are $z_1 = \frac{1+\sqrt{5}}{2}$ and $z_2 = \frac{1-\sqrt{5}}{2}$. We now apply the bilinear transform and check this result. Replacing the bilinear transform in $p(z)$ we obtain

$$z^2 - z - 1 = \left(\frac{1+s}{1-s}\right)^2 - \left(\frac{1+s}{1-s}\right) - 1 = \frac{s^2 + 4s - 1}{(1-s)^2} \tag{14.25}$$

By direct calculation of its roots or by the inspection of the signs of its coefficients, we immediately conclude that one of the roots of $p(s) = s^2 + 4s - 1$ is negative and the other is positive, thus confirming the result of the analysis of $p(z)$.

Example 14.3

Let us now consider a system with a transfer function whose denominator is $p(z) = 4z^3 + 3z^2 + 2z + 1$. This denominator has all roots inside the unit circle in virtue of the following fact: given a polynomial $p(z) = \varphi_0 z^n + \varphi_1 z^{n-1} + \ldots + \varphi_{n-1} z + \varphi_n$, then a sufficient condition for the asymptotic stability is that $\varphi_0 > \varphi_1 > \ldots > \varphi_n > 0$. We now apply the bilinear transform and check this result. Substituting the bilinear transform in $p(z)$ we obtain

$$
\begin{aligned}
4z^3 + 3z^2 + 2z + 1 &= 4\left(\tfrac{1+s}{1-s}\right)^3 + 3\left(\tfrac{1+s}{1-s}\right)^2 + 2\left(\tfrac{1+s}{1-s}\right) + 1 \\
&= \tfrac{4(1+s)^3 + 3(1+s)^2(1-s) + 2(1+s)(1-s)^2 + (1-s)^3}{(1-s)^3} \\
&= \tfrac{2s^3 + 10s^2 + 10s + 10}{(1-s)^3}
\end{aligned} \tag{14.26}
$$

The Routh table of $p(s) = 2s^3 + 10s^2 + 10s + 10$ is given in Table 14.2, from which we conclude that the continuous-time system, and, so, also the discrete-time one, is asymptotically stable.

TABLE 14.2
Routh table of $p(s) = 2s^3 + 10s^2 + 10s + 10$.

3	2	10
2	10	10
1	8	
0	10	

14.6 Digital control systems

In modern control engineering practice, digital controllers are frequently used. This implies that the continuous-time system to control is managed by a discrete-time system, typically a microcontroller. There are two control diagrams that represent this scenario, as shown in Figs. 14.2 and 14.3. Here, we notice the presence of two new blocks, A/D, which stands for the analog-to-digital converter, and D/A, which stands for the digital-to-analog converter. They are fundamental to interface the controller, which is a digital device, with the other blocks that operate with analog signals in continuous time. The difference between the two diagrams of Figs. 14.2 and 14.3 that, in the first case, the error is an analog signal, while, in the second case, the error is a digital signal. In the first case, an A/D converter is placed just after the adder that computes the error, while in the second case the A/D converter is placed after the sensor. In addition, the reference is provided as a digital signal or it is transformed into a digital signal by another A/D.

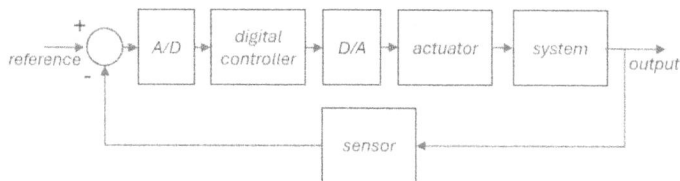

FIGURE 14.2
Scheme of a feedback digital control system.

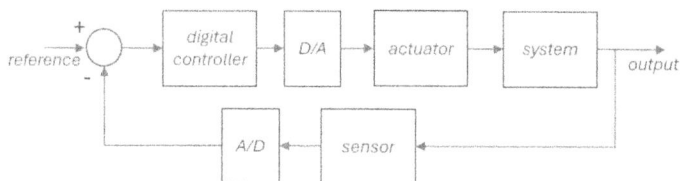

FIGURE 14.3
Scheme of a feedback digital control system where the error signal is digital.

The purpose of the A/D converter is to transform the analog input into a digital output. This process involves two operations. The first is sampling. The input signal is typically sampled at regular intervals of time, called sampling time T_s. The second operation performed by the A/D is to encode each sampled value in a string of bits. Each A/D device is thus associated with a sampling period or its inverse, the sampling frequency, and with a number of bits that determines the accuracy of encoding.

The D/A converter transforms the digital input into an analog signal. The simplest way to perform this operation is the so-called zero-order holder (ZOH), which holds the value of a sample for all time up to the next sample, such that the corresponding analog signal is $x_{ZOH}(t) = x_h$ for $hT_s \leq t < (h+1)T_s$.

The use of digital controllers is widespread as they offer several advantages. They provide accurate calculations and offer high computational capabilities. For this reason, they can handle from simple to more complex control algorithms, hence offering a large flexibility in terms of the control law that can be implemented. These control schemes are also easy to expand to incorporate new devices and functionalities. All signals can be easily tracked and recorded for real-time monitoring, fault diagnosis, and post-processing analysis of the system behavior.

On the other hand, digital control systems are more difficult to design. For example, the specifications on the z-plane are more difficult to satisfy than their continuous-time counterpart.

In addition, the digital control scheme introduces a time delay equal to $T_s/2$. Small values of this time delay can be neglected, but, when this is in the same order of magnitude of the time scales of the system to control, one has to explicitly take it into account.

Although there are techniques to design the control law tailored to the discrete-time case, an efficient way to design a digital control law is to design the continuous-time controller using the techniques available for the continuous-time case and considering that the time delay of the system to control is the open-loop one, if any, plus a further delay equal to $T_s/2$, due to the use of the digital technology. Once obtained the continuous-time controller, e.g., the transfer function $C(s)$, then the corresponding digital control law is derived using several techniques available to the purpose. In Lecture 17 we will show an example of a compensator designed using the procedure for continuous-time systems and, then, transformed into a discrete-time system.

14.7 FIR and IIR systems

In this section, we focus on two particular classes of discrete-time systems which have relevance in digital filter design.

14.7.1 Finite impulse response systems

In the previous lecture, we noted that discrete-time systems with all poles at the origin exhibit the peculiar behavior of having modes that reach the exact value of zero after a finite number of samples, k, where k is the multiplicity of the pole at the origin. These systems are known as Finite Impulse Response (FIR) systems, as, clearly, the same behavior is observed for the impulse response that becomes zero after k samples.

In the z-domain, a FIR system is characterized by a transfer function that can be written as follows:

$$\frac{Y(z)}{U(z)} = G(z) = b_0 + \frac{b_1}{z} + \ldots + \frac{b_n}{z^n} = \frac{b_0 z^n + b_1 z^{n-1} + \ldots + b_{n-1} z + b_n}{z^n}$$

$$(14.27)$$

Being $U(z) = 1$ the z-transform of the impulse, the samples of the impulse response are indeed $b_0, b_1, \ldots, b_n$.

Example 14.4 _____

Let us consider the discrete-time system with a transfer function

$$G(z) = \frac{3z^4 + 0.1z^3 + z^2 + 0.2z + 2}{z^4} \tag{14.28}$$

This is a FIR system as all poles are at the origin. Considering $G(z) = Y(z)/U(z)$, we get that

$$z^4 Y(z) = (3z^4 + 0.1z^3 + z^2 + 0.2z + 2)U(z) \tag{14.29}$$

and applying the inverse z-transform we find that

$$y(h) = 3u(h) + 0.1u(h-1) + u(h-2) + 0.2u(h-3) + 2u(h-4) \tag{14.30}$$

Notice that the response depends solely on the input signal and its regressions. This response can be numerically calculated in MATLAB® by using the following commands:

```
z=tf('z',1)
G=(3*z^4+0.1*z^3+z^2+0.2*z+2)/z^4;
impulse(G)
```

The result is shown in Fig. 14.4.

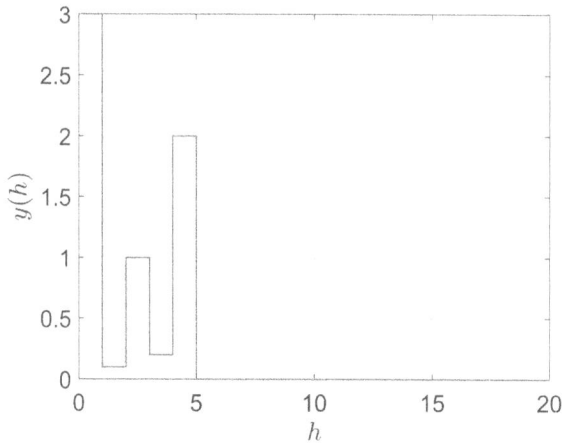

FIGURE 14.4
Impulse response of the FIR system with transfer function (14.28).

14.7.2   Infinite impulse response systems

A discrete-time system whose impulse response is characterized by an infinite sequence of infinite samples is called Infinite Impulse Response (IIR) system. The transfer function of an IIR system reads

$$G(z) = \frac{\displaystyle\sum_{i=0}^{m} b_i z^{-i}}{\displaystyle\sum_{j=0}^{n} a_j z^{-j}} \tag{14.31}$$

that corresponds to the finite-difference iteration

$$y(h) = \frac{1}{a_0}\left(\sum_{i=0}^{m} b_i u(h-i) - \sum_{j=1}^{n} a_j y(h-j)\right) \tag{14.32}$$

In this case, the output depends on some previous values of the input and of the output itself.

Example 14.5 _____

Let us consider the system with the transfer function

$$G(z) = \frac{z^{-1} + 2z^{-2} + z^{-3} + z^{-4}}{1 + 0.7z^{-1} - 0.24z^{-2} - 0.148z^{-3} - 0.016z^{-4}} \tag{14.33}$$

This is an IIR system whose output can be obtained as follows:

$$\begin{aligned} y(h) &= u(h-1) + 2u(h-2) + u(h-3) + u(h-4) - 0.7y(h-1) \\ &\quad + 0.24y(h-2) + 0.148y(h-3) + 0.016y(h-4) \end{aligned} \tag{14.34}$$

The impulse response can be calculated with the following MATLAB® commands:

```
z=tf('z',1)
G=(z^-1+2*z^-2+z^-3+z^-4)/(1+0.7*z^-1-0.24*z^-2-0.148*z^-3-0.016*z^-4)
impulse(G)
```

or, equivalently

```
u=zeros(100,1);
y=zeros(100,1);
u(4)=1;
for h=5:100
y(h)=(u(h-1)+2*u(h-2)+u(h-3)+u(h-4)+0.016*y(h-4)+0.148*y(h-3)...
+0.24*y(h-2)-0.7*y(h-1));
end
stairs(y(4:end));
```

The result is illustrated in Fig. 14.5, which shows that zero is approached asymptotically and not in a finite number of steps.

Note that, when the finite-difference equation is iterated, the impulse is applied to the fourth element of the input vector, such that the first element that is computed corresponds to $h = 5$, consistently with the fourth-order regression needed to compute $y(h)$.

14.7.3   FIR approximation of IIR systems

FIR systems are particularly useful in approximating the dynamics of asymptotically stable IIR systems. The response $y(h)$ of an asymptotically stable IIR system, in fact, tends to zero for $h \to \infty$. However, for a sufficiently high value of h, the response is vary close to zero. Specifically, let $y(h) \simeq 0$ for $h > \Theta$. Under this assumption, the IIR system can be approximated with the transfer function of a FIR system of order Θ as follows:

$$\tilde{G}(z) = \frac{y(0)z^{\Theta} + y(1)z^{\Theta-1} + \ldots + y(\Theta)}{z^{\Theta}} \tag{14.35}$$

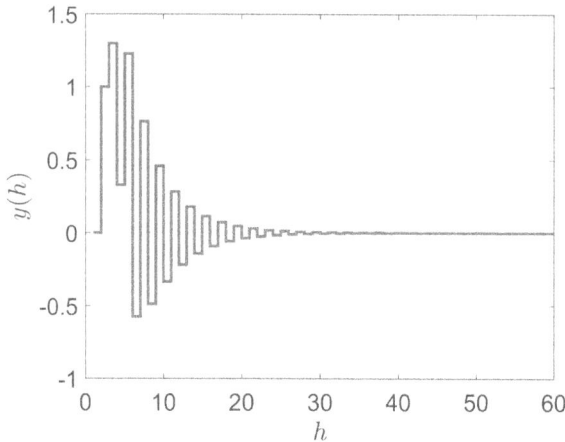

FIGURE 14.5
Impulse response of the IIR system with transfer function (14.33).

The FIR approximation can be of order significantly higher than the order of the original IIR. This technique, however, allows to obtain an approximated frequency response starting from the knowledge of the samples of the impulse response, without knowing the IIR system dynamics.

Example 14.6 _____

Let us consider the IIR system with transfer function

$$G(z) = \frac{z - 0.2}{z - 0.5} \tag{14.36}$$

The impulse response, shown in Fig. 14.6, is approximately zero for $h > 7$, therefore a FIR approximation can be given by

$$\tilde{G}(z) = 1.003\frac{z^7 + 0.3z^6 + 0.15z^5 + 0.075z^4 + 0.0375z^3 + 0.0187z^2 + 0.00937z + 0.00469}{z^7} \tag{14.37}$$

where the coefficients of the numerator polynomial are the first seven samples of the impulse response of $G(z)$, and the gain 1.003 ensures that the two static gains match, i.e. $\tilde{G}(1) = G(1)$. The impulse response of $\tilde{G}$ is exactly zero after seven samples. For the purpose of comparison, the step responses of the IIR system and its FIR approximation are both illustrated in Fig. 14.7.

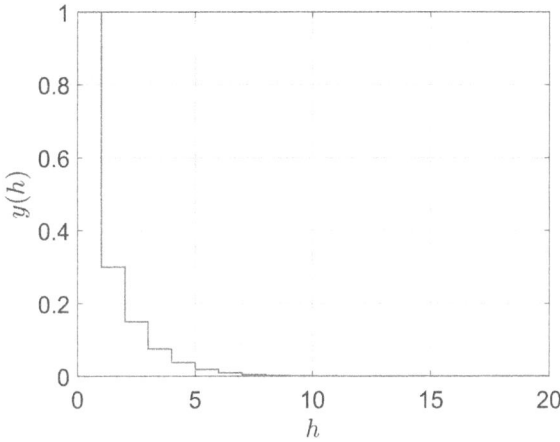

FIGURE 14.6
Impulse response of the IIR system with transfer function (14.36).

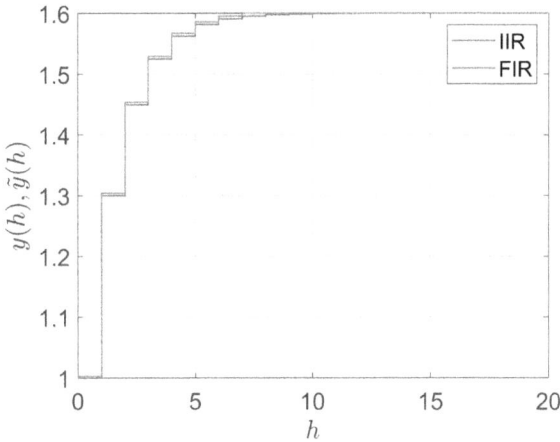

FIGURE 14.7
Step responses of the IIR system with transfer function (14.36) and of its FIR approximation with transfer function (14.37).

14.8   Exercises

1. Calculate the z-transform of the discrete-time signal $x(h) = 2 - h4^{h-1}$.

2. Calculate the z-transform of the discrete-time signal $x(h) = 2^{h-1}$.

3. Calculate the inverse z-transform of $X(z) = \frac{z}{z^2-3z-4}$.

4. Determine the impulse response of the discrete-time system $G(z) = \frac{0.2z^5-z^4+2z^3+3z^2-z+1}{z^5}$.

5. Determine a suitable FIR approximation of the discrete-time system $G(z) = \frac{z^2-0.5z+0.2}{z^2+0.5z+0.2}$

15

Lecture 15—Controller design: steady-state performance

In this lecture, we first introduce the sensitivity functions. These are the transfer functions that fully characterize the input-output relationships in a feedback control scheme with unity feedback. Next, we introduce the notion of system type, which is fundamental to analyze the steady-state behavior of the system in response to some canonical inputs, such as the step, the ramp and the parabolic input. Finally, we will discuss how to design a compensator to satisfy specifications on the steady-state behavior of a closed-loop system.

15.1   Control characteristics

In this and the next two lectures, we will refer to the control scheme with unity feedback shown in Fig. 15.1. Beyond the reference signal $r(t)$, there are two other inputs in the system, the disturbances $d(t)$ and $n(t)$. $d(t)$ represents an additive disturbance acting on the system output $y(t)$, while $n(t)$ an additive disturbance directly affecting the error $e(t)$. It can therefore represent the measurement error incurred while measuring $y(t)$. We will see later on that the effect of these two disturbances on $y(t)$, and ultimately on the performance of the whole system, is very different.

Key variables of this control system are the output $y(t)$, but also the control signal $u(t)$ and the error $e(t)$. More specifically, studying the time evolution of $u(t)$ is fundamental to understand whether the planned control action is moderate or not. A control action that is too large could move the system away from the linearity assumption and the operating conditions for which the system model was derived, potentially rendering the control action ineffective in practice.

Any controller must achieve:

- system stability;

- tracking of the output;

DOI: 10.1201/9781003487289-15

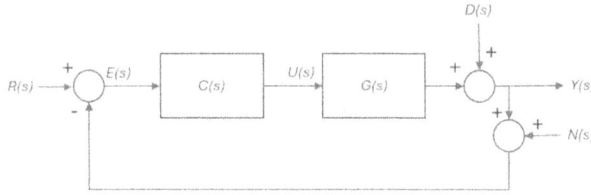

FIGURE 15.1
Control scheme with unity feedback.

- disturbance rejection;

- robustness to changes in the system parameters.

Stability of the closed-loop system is the primary and most important objective of any controller. Without closed-loop stability, the other specifications for the control system become meaningless. As an example, suppose we want to control the temperature of a room such that it settles to a value of 21°C. If the closed-loop system is unstable, it will not reach any equilibrium, making it impossible to achieve the control objective.

Once we have ensured that the closed-loop system is stable, we can require the controller to enable the system to track the input. This allows the output to follow a desired time evolution. For example, one can specify a velocity profile for a motor and ask the controller to ensure that the output tracks this profile.

The closed-loop system must be able to eliminate or, at least, mitigate the effects of disturbances, as these can drive the system away from the desired behavior. This property is called disturbance rejection.

Lastly, since the system model is subject to modeling errors or time-varying parameters, the controller is also required to be robust to changes in these parameters. This means that, if the parameters deviate from their nominal values, the controller must still be able to ensure performance similar to those achieved with the nominal parameters. Typically, this requirement cannot be satisfied for all possible deviations but should be met within specified ranges of parameter uncertainty.

15.2   Sensitivity functions

We define the following transfer functions:

- *sensitivity* $S(s)$

$$S(s) = \frac{1}{1 + C(s)G(s)} \tag{15.1}$$

- *complementary sensitivity $F(s)$*

$$F(s) = \frac{C(s)G(s)}{1 + C(s)G(s)} \tag{15.2}$$

- *control sensitivity $Q(s)$*

$$Q(s) = \frac{C(s)}{1 + C(s)G(s)} \tag{15.3}$$

These functions are fundamental, as they allow to fully characterize the behavior of the control system in Fig. 15.1. More specifically, they are crucial to determine the input-output relationships between the inputs $r(t)$, $d(t)$, and $n(t)$, and the key variables $y(t)$, $u(t)$, and $e(t)$. Consider, for example, the input-output relationship from input $r(t)$ to output $y(t)$. It can be obtained by applying the superposition principle to the control system in Fig. 15.1. This means to consider $d(t) = 0$ and $n(t) = 0$. Under this assumption, the equivalent diagram of Fig. 15.2(a) is obtained, from which we derive that:

$$\frac{Y(s)}{R(s)} = \frac{C(s)G(s)}{1 + C(s)G(s)} = F(s) \tag{15.4}$$

Next, consider the input-output relationship from input $r(t)$ to the feedback error $e(t)$. The superposition principle with $d(t) = 0$ and $n(t) = 0$ yields the equivalent diagram of Fig. 15.2(b), from which we get:

$$\frac{E(s)}{R(s)} = \frac{1}{1 + C(s)G(s)} = S(s) \tag{15.5}$$

The sensitivity $S(s)$ also determines the way in which the disturbance $d(t)$ affects the output $y(t)$. In fact, by setting $r(t) = 0$ and $n(t) = 0$, we obtain the equivalent diagram of Fig. 15.2(c), from which we derive that:

$$\frac{Y(s)}{D(s)} = \frac{1}{1 + C(s)G(s)} = S(s) \tag{15.6}$$

Instead, the input-output relationship from $n(t)$ to $y(t)$ is obtained by considering $r(t) = 0$ and $d(t) = 0$. The equivalent diagram of Fig. 15.2(d) yields that:

$$\frac{Y(s)}{N(s)} = -\frac{C(s)G(s)}{1 + C(s)G(s)} = -F(s) \tag{15.7}$$

With similar arguments the relationships between all other inputs and outputs can be derived. In summary, we have the following:

$$\begin{bmatrix} Y(s) \\ U(s) \\ E(s) \end{bmatrix} = \begin{bmatrix} F(s) & S(s) & -F(s) \\ Q(s) & -Q(s) & -Q(s) \\ S(s) & -S(s) & F(s) \end{bmatrix} \begin{bmatrix} R(s) \\ D(s) \\ N(s) \end{bmatrix} \tag{15.8}$$

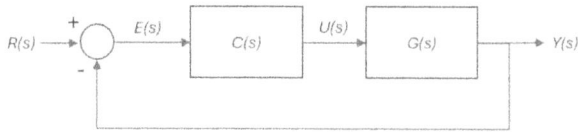

(a)

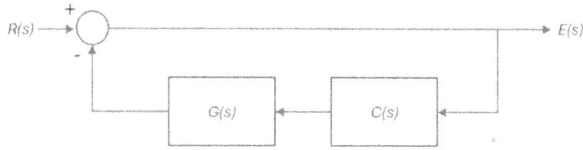

(b)

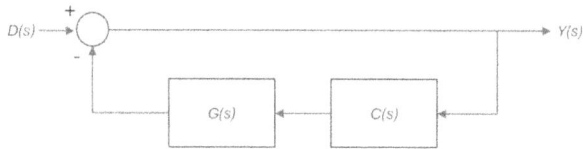

(c)

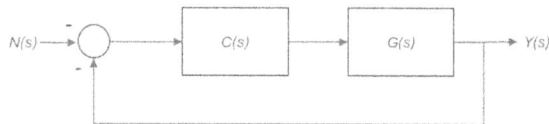

(d)

FIGURE 15.2
Equivalent diagrams for the derivation of the input-output relationships of
Fig. 15.1, obtained by setting: (a) $d(t) = 0$ and $n(t) = 0$, input $r(t)$, output
$y(t)$; (b) $d(t) = 0$ and $n(t) = 0$, input $r(t)$, output $e(t)$; (c) $r(t) = 0$ and
$n(t) = 0$, input $d(t)$, output $y(t)$; (d) $r(t) = 0$ and $d(t) = 0$, input $n(t)$, output
$y(t)$.

We note that the three sensitivity functions are not independent of each other but are interrelated. In fact, we have that $F(s) + S(s) = 1$ and $Q(s) = F(s)G^{-1}(s) = C(s)S(s)$.

Assuming the closed-loop system is asymptotically stable, control performance requirements can be directly translated into requirements for the sensitivity functions. However, due to the interdependence of the sensitivity functions, satisfying some requirements may conflict with satisfying others, necessitating trade-off solutions.

The ability of the output $y(t)$ to track the input $r(t)$ would ideally require $F(s) = 1$ for any s. Moreover, for the disturbance $d(t)$ to have no effect on the output $y(t)$, it would ideally require $S(s) = 0$ for any s. These two requirements are compatible with each other, but conflict with the fact that $Y(s)/N(s) = -F(s)$. Thus, having $F(s) = 1$ would result in no rejection of the effect of $n(t)$ on $y(t)$.

Furthermore, since $Q(s) = G^{-1}(s)$ if $F(s) = 1$, and $G(s)$ is a strictly proper function, $Q(s)$ would have a large magnitude at high frequencies. This results in large-amplitude high-frequency components in the control signal $u(t)$, which contrasts with the need for a moderate control action.

In conclusion, to satisfy these conflicting requirements, it is appropriate to find a trade-off in the frequency range where $F(j\omega)$ should be close to one, facilitating the tracking of the input $r(t)$ and the rejection of disturbances $d(t)$ with frequency components in this range. Conversely, $F(j\omega)$ should be close to zero in the region where it is important to promote the rejection of disturbances $n(t)$ within that frequency range.

15.3 System type

With reference to the unity feedback configuration, let us rewrite the open-loop transfer function as follows:

$$C(s)G(s) = \frac{K_m C'(s)G'(s)}{s^m} \tag{15.9}$$

where $C'(s)$ and $G'(s)$ are such that $C'(0) = 1$ and $G'(0) = 1$, that is, they do not have poles at the origin and have gain equal to one. The integer m indicates how many poles at the origin are present in the cascade of $C(s)$ and $G(s)$. This is known as the *system type*. Typical cases are $m = 0, 1, 2$, namely type 0 systems, type 1 systems, and type 2 systems. The gain K_m takes different names as m varies. For $m = 0$, $K_m = K_p$ is known as the position gain; for $m = 1$ $K_m = K_v$ is known as the velocity gain; and for $m = 2$ $K_m = K_a$ is known as the acceleration gain. These are historical terms originating from the early applications in the field of position control.

Here, we study the steady-state behavior of the closed-loop system. Specifically, we examine whether the closed-loop system can asymptotically track step, ramp, or parabolic inputs. Therefore, we consider $r(t) = R_k \frac{t^k}{k!}\text{step}(t)$ with R_k denoting the amplitude of the signal. In this way, for $k = 0$, we get $r(t) = R_0\text{step}(t)$, namely the step input; for $k = 1$, we have $r(t) = R_1 t\text{step}(t)$, namely the ramp input; and, for $k = 2$, we obtain $r(t) = R_2 \frac{t^2}{2}\text{step}(t)$, namely the parabolic input. The Laplace transform of these functions is $R(s) = \frac{R_k}{s^{k+1}}$.

We calculate the steady-state behavior of the tracking error $e(t) = r(t) - y(t)$ using the transfer function from input $R(s)$ to output $E(s)$, namely the complementary sensitivity function. Thus, we have

$$E(s) = \frac{1}{1 + C(s)G(s)}R(s) \tag{15.10}$$

Assuming that the closed-loop system is asymptotically stable, then the steady-state value of $e(t)$, denoted as e_{ss}, is calculated by applying the final value theorem. This assumption is entirely realistic, as ensuring stability is the primary requirement of control. Doing so, we get

$$
\begin{aligned}
e_{ss} &= \lim_{t \to +\infty} e(t) = \lim_{s \to 0} sE(s) = \lim_{s \to 0} s\frac{1}{1+C(s)G(s)} \frac{R_k}{s^{k+1}} \\
&= \lim_{s \to 0} \frac{s^m}{s^m + K_m} \frac{R_k}{s^k}
\end{aligned} \tag{15.11}
$$

Using Eq. (15.11), let us now evaluate e_{ss} for type 0, type 1, and type 2 systems with step, ramp, and parabolic inputs.

Type 0 systems. Type 0 systems have no poles at the origin, $m = 0$. For the step input, where $k = 0$, we have $e_{ss} = \frac{R_0}{1+K_p}$. In fact, the output converges to the steady-state value $y_{ss} = \frac{K_p R_0}{1+K_p}$, as it can be obtained considering that $y_{ss} = \lim_{t \to +\infty} y(t) = \lim_{s \to 0} s\frac{C(s)G(s)}{1+C(s)G(s)} \frac{R_0}{s}$. Hence, the steady-state output is also constant, but at a value lower than the input (assumed $K_p > 0$).

Let us now consider a ramp signal, i.e., $k = 1$. In this case $e_{ss} = \infty$, indicating that the ramp function cannot be followed by the system. In fact, in such case, expanding $Y(s)$ in simple fraction we find that the residue associated with the term $1/s^2$ is given by

$$\lim_{s \to 0} s^2 \frac{C(s)G(s)}{1+C(s)G(s)} \frac{R_1}{s^2} = \frac{K_p R_1}{1+K_p} \tag{15.12}$$

that suggests that the system output has a ramp term with a smaller slope than the input, such that the input cannot be tracked.

Lastly, let us consider the parabolic input, $k = 2$. Also in this case, Eq. (15.11) gives $e_{ss} = \infty$, such that the parabolic input cannot be followed by type 0 systems.

Type 1 systems. Type 1 systems have one pole at the origin, $m = 1$. For step inputs, $k = 0$. In this case, Eq. (15.11) gives $e_{ss} = 0$, indicating that the step input is tracked with zero error. This is also confirmed by the calculation of the steady-state value of the output, namely $y_{ss} = \lim_{t \to +\infty} y(t) =$ $\lim_{s \to 0} s \frac{C(s)G(s)}{1+C(s)G(s)} \frac{R_0}{s} = \lim_{s \to 0} s \frac{K_v}{s+K_v} \frac{R_0}{s} = R_0$. The final value of the output corresponds to the amplitude of the input signal, hence $e_{ss} = 0$.

Let us now move to the ramp input, i.e., $k = 1$. In this case, from Eq. (15.11) we derive that $e_{ss} = \frac{R_1}{K_v}$. This means that, after the transient, the output is characterized by a ramp term with a slope equal to that of the input. The difference between output and input remains constant. Once again, this behavior can be explained by expanding $Y(s)$ in simple fractions. The residue associated with the term $1/s^2$ is given by

$$\lim_{s \to 0} s^2 \frac{C(s)G(s)}{1+C(s)G(s)} \frac{R_1}{s^2} = \lim_{s \to 0} \frac{K_v}{s+K_v} R_1 = R_1 \tag{15.13}$$

indicating that input and output have the same slope. However, $y(t)$ contains another term that does not vanish after the transient. This is the term that in the expansion in simple fractions of $Y(s)$ is associated with $1/s$, and it is the reason for a steady-state constant error between input and output.

Lastly, for parabolic inputs, $k = 2$, type 1 systems have $e_{ss} = \infty$, indicating that these systems cannot track such signals.

Type 2 systems. Type 2 systems have two poles at the origin, $m = 2$. For both step ($k = 0$) and ramp ($k = 1$) inputs, Eq. (15.11) gives $e_{ss} = 0$. Therefore, type 2 systems can track both step and ramp inputs with zero steady-state error.

For parabolic inputs, $k = 2$, from Eq. (15.11) we derive that $e_{ss} = \frac{R_2}{K_a}$, such that the error is constant but non-zero.

The results for the three types of systems are summarized in Table 15.1.

TABLE 15.1
Steady-state error e_{ss} (Eq. (15.11)) for type 0, type 1, and type 2 systems with step, ramp, and parabolic inputs.

system type	step	ramp	parabolic input
type 0	$\frac{R_0}{1+K_p}$	∞	∞
type 1	0	$\frac{R_1}{K_v}$	∞
type 2	0	0	$\frac{R_2}{K_a}$

15.4   Controller design for steady-state error requirements

The analysis on the system type with the help of Table 15.1 guides the selection of the transfer function $C(s)$ to satisfy one or more requirements for the tracking capabilities of a control system. It is worth noticing that the system type counts the number of poles at the origin of the transfer function $C(s)G(s)$. This means that if, for instance, a zero steady-state error to a step input is required, then it is sufficient for either the process or the controller to have a pole at the origin. Specifically, if the process does not meet this requirement, a controller with a pole at the origin should be designed.

The design of $C(s)$ involves four steps:

1. The minimum system type that meets all given requirements is determined; let this number be m.

2. The gain K_m is adjusted to satisfy any additional criteria.

3. Let μ be the static gain and r the number of poles at the origin of $G(s)$. $C(s)$ is selected so as to have at least $m - r$ poles at the origin, and a gain larger than or equal to K_m/μ.

4. The closed-loop stability is checked.

This procedure is now illustrated with a series of examples.

Example 15.1 _____

Given $G(s) = \frac{1}{(s+1)(s+2)}$, determine a controller $C(s)$ such that $e_{ss} = 0$ for a unit step input and $|e_{ss}| \leq 0.5$ for a unit ramp input.
To satisfy the requirement that $e_{ss} = 0$ for a step input, $C(s)G(s)$ must be at least type 1, i.e., $m = 1$. Since $G(s)$ has no pole at the origin, a pole at the origin must be introduced in $C(s)$. Hence, we set $C(s) = K_c/s$. To meet the requirement that $|e_{ss}| \leq 0.5$ for a ramp input, since in this case $e_{ss} = 1/K_v$ (see Table 15.1), then $K_v \geq 2$. Since $K_v = K_c/2$, then $K_c \geq 4$. Next, we calculate the closed-loop transfer function $F(s) = \frac{C(s)G(s)}{1+C(s)G(s)}$:

$$F(s) = \frac{K_c}{s^3 + 3s^2 + 2s + K_c} \tag{15.14}$$

and apply the Routh criterion to the denominator. The results are illustrated in Table 15.2, which gives that for stability $0 < K_c < 6$. Hence, to meet the requirements of the control, we can select $4 \leq K_c < 6$.

TABLE 15.2
Routh table for $p(s, K_c) = s^3 + 3s^2 + 2s + K_c$.

3	1	2
2	3	K_c
1	$\frac{6-K_c}{3}$	
0	K_c	

Example 15.2

Given $G(s) = \frac{1}{s^2+s+1}$, determine a controller $C(s)$ such that $e_{ss} = 0$ for a unit step input and $|e_{ss}| \leq 0.2$ for a unit ramp input.

In order to have $e_{ss} = 0$ for a step input, $C(s)G(s)$ must have at least a pole at the origin, and, as $G(s)$ does not meet this requirement, then a pole at the origin must be introduced in $C(s)$. We fix $C(s) = K_v/s$. To have $|e_{ss}| \leq 0.2$ for a ramp input, $K_v \geq 5$ must hold. From the analysis of the stability of the closed-loop transfer function $F(s) = \frac{C(s)G(s)}{1+C(s)G(s)}$:

$$F(s) = \frac{K_v}{s^3 + s^2 + s + K_v} \tag{15.15}$$

we get that $0 < K_v < 1$ (see Table 15.3). The requirement $K_v \geq 5$ conflicts with the stability condition, and therefore cannot be satisfied. In the next lecture we will show that this problem can be solved by introducing in $C(s)$ a zero and a pole in a configuration that does not affect the steady-state performance.

TABLE 15.3

Routh table for $p(s, K_v) = s^3 + s^2 + s + K_v$.

3	1	1
2	1	K_v
1	$1 - K_v$	
0	K_v	

Example 15.3

Given $G(s) = \frac{s+0.1}{s^2+s}$, determine a controller $C(s)$ such that $y_{ss} = 0$ for a unit step disturbance and $|y_{ss}| \leq 0.05$ for a unit ramp disturbance.

For the first requirement ($y_{ss} = 0$ for a step disturbance) $C(s)G(s)$ must be at least type 1. Since $G(s)$ has a pole at the origin, then $C(s) = K_c$ can be selected. To have $|y_{ss}| \leq 0.05$ for a ramp disturbance, then $K_v \geq 20$. Since $K_v = 0.1K_c$, then $K_c \geq 200$. Closed-loop stability is analyzed by calculating the sensitivity function $S(s) = \frac{1}{1+C(s)G(s)}$:

$$S(s) = \frac{s(s+1)}{s^2 + (K_c+1)s + 0.1K_c} \tag{15.16}$$

from which we immediately get that $K_c > 0$. Therefore, selecting $C(s) = K_c$ with $K_c \geq 200$ solves the problem.

Example 15.4

Given $G(s) = \frac{s+0.1}{s^2+s}$, determine a controller $C(s)$ such that $y_{ss} = 0$ for a unit step disturbance and $y_{ss} = 0$ for a unit ramp disturbance.

To meet the two requirements, $C(s)G(s)$ must be at least type 2. Since $G(s)$ has a pole at the origin, the second pole at the origin must be provided by $C(s) = K_c/s$. Here $K_a = 0.1K_c$. The value of K_c is selected based on the conditions for closed-loop stability. Specifically, we calculate the sensitivity function $S(s) = \frac{1}{1+C(s)G(s)}$, obtaining

$$S(s) = \frac{s^2(s+1)}{s^3 + s^2 + K_c s + 0.1K_c} \tag{15.17}$$

From the Routh Table 15.4 of $p(s, K_a) = s^3 + s^2 + K_c s + 0.1K_c$ we derive that $K_c > 0$ guarantees stability. Hence, any $C(s)$ with $K_c > 0$ solves the problem.

TABLE 15.4
Routh table for $p(s, K_a) = s^3 + s^2 + K_c s + 0.1K_c$.

3	1	1
2	1	$0.1K_c$
1	$0.9K_c$	
0	$0.1K_c$	

The compensator designed for steady-state performance consists of a gain and additional poles at the origin. Generally, increasing the system type may lead to instability. As seen in Example 15.2, there are scenarios where this type of compensator cannot be used alone, as it may fail to meet the primary requirement of stability. In such cases, a controller that includes one or more pairs of poles and zeros may address the issue. This type of controller also affects the dynamic performance of the system during transients and will be the subject of the next lecture.

15.5   MATLAB examples

In this section, using the sensitivity functions studied in this Lecture, we calculate in MATLAB® the time evolution of $y(t)$ for the examples discussed in the previous section, checking that the designed controllers satisfy the steady-state performance.

MATLAB® exercise 15.1 _____

Let us consider Example 15.1 of Sec. 15.4, where we have found that the system $G(s) = \frac{1}{(s+1)(s+2)}$ the controller $C(s) = \frac{K_c}{s}$ with $4 \leq K_c < 6$ can guarantee $e_{ss} = 0$ for a unit step input and $|e_{ss}| \leq 0.5$ for a unit ramp input. Let us now fix $K_c = 4$ and check with MATLAB® that the controller achieves the desired steady-state performance.
First, we define the relevant transfer functions, namely $G(s)$, $C(s)$, and $F(s)$ with the commands:

```
s=tf('s');
G=1/((s+1)*(s+2));
Kc=4;
C=Kc/s;
F=feedback(C*G,1);
```

Then, we compute the step response and plot the result:

```
[Y,T]=step(F,100);
figure,plot(T,ones(length(T),1),T,Y)
```

Next, we compute the response to the unit ramp and plot the result:

```
t=0:0.01:100;
ramp=t;
[Y,T]=lsim(F,ramp,t);
figure,plot(T,ramp,T,Y)
```

The time evolution of $y(t)$, along with the input $r(t)$, is shown in Fig. 15.1. When $r(t) = \text{step}(t)$, the output $y(t)$ converges to 1, such that the steady-state error is exactly zero (Fig. 15.3(a)). When $r(t) = t$, after a transient the output $y(t)$ grows linearly with t; the steady-state error is constant at a value equal to -0.5.

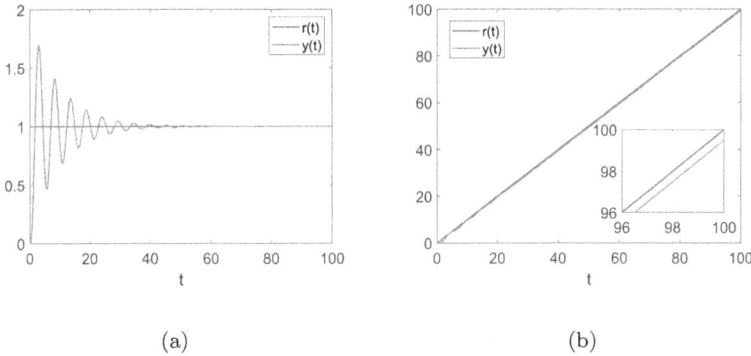

(a) (b)

FIGURE 15.3

Time evolution of the output $y(t)$ for the closed-loop system with $G(s) = \frac{1}{(s+2)(s+5)}$ and $C(s) = \frac{10}{s}$: (a) response to the unit step; (b) response to the unit ramp.

MATLAB® exercise 15.2 _____

Let us consider Example 15.3 of Sec. 15.4, where we have found that for the system $G(s) = \frac{s+0.1}{s(s+1)}$ the controller $C(s) = K_c$ with $K_c \geq 200$ can guarantee $y_{ss} = 0$ for a unit step disturbance and $|y_{ss}| \leq 0.05$ for a unit ramp disturbance. Let us now fix $K_c = 200$ and check with MATLAB® that the controller achieves the desired steady-state performance.
In this case the relevant transfer functions are $G(s)$, $C(s)$, and $S(s)$:

```
s=tf('s');
G=(s+0.1)/(s*(s+1));
Kc=200;
C=Kc;
S=feedback(1,C*G);
```

Then, we compute the step response and plot the result:

```
[Y,T]=step(S,0.2);
figure,plot(T,Y)
```

Next, we compute the response to the unit ramp and plot the result:

```
t=0:0.001:100;
ramp=t;
[Y,T]=lsim(S,ramp,t);
figure,plot(T,Y)
```

The time evolution of $y(t)$ for the two disturbances is shown in Fig. 15.2. When $d(t) = \text{step}(t)$, we find that the output $y(t)$ converges to zero, indicating that the effect of the disturbance asymptotically vanishes (Fig. 15.4(a)). When $d(t) = t$, after a transient the output $y(t)$ converges to the value 0.05, indicating that also in this case the controller meets the given requirement on disturbance rejection.

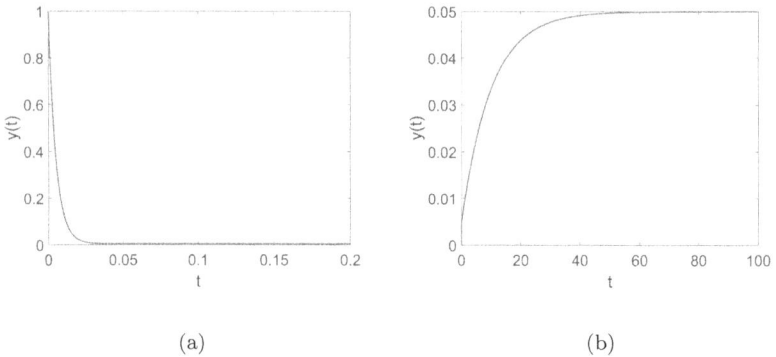

(a)   (b)

FIGURE 15.4

Time evolution of the output $y(t)$ for the closed-loop system with $G(s) = \frac{s+0.1}{s(s+1)}$ and $C(s) = 200$ in response to a disturbance: (a) response to the unit step disturbance; (b) response to the unit ramp disturbance.

MATLAB® exercise 15.3 ───

Let us consider Example 15.4 of Sec. 15.4, where we have found that for the system $G(s) = \frac{s+0.1}{s(s+1)}$ the controller $C(s) = K_c/s$ with $K_c > 0$ can guarantee $y_{ss} = 0$ for a unit step disturbance and a unit ramp disturbance. Let us now fix $K_c = 20$ and check with MATLAB® that the controller achieves the desired steady-state performance. The relevant transfer functions are $G(s)$, $C(s)$, and $S(s)$:

```
s=tf('s');
G=(s+0.1)/(s*(s+1));
Kc=20;
C=Kc/s;
S=feedback(1,C*G);
```

Then, we compute the step response and plot the result:

```
[Y,T]=step(S,50);
figure,plot(T,Y)
```

The response to the unit ramp is calculated and plotted using the commands:

```
t=0:0.001:50;
ramp=t;
[Y,T]=lsim(S,ramp,t);
figure,plot(T,Y)
```

The time evolution of $y(t)$ for the two disturbances is shown in Fig. 15.3. In both cases, after a transient the output vanishes, indicating that the controller is able to reject the two types of disturbances.

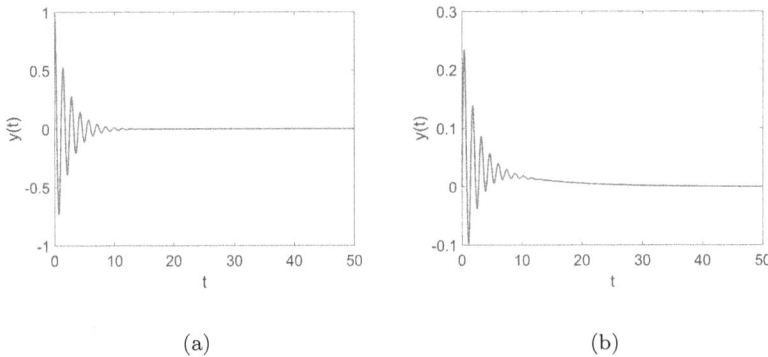

(a)   (b)

FIGURE 15.5
Time evolution of the output $y(t)$ for the closed-loop system with $G(s) = \frac{s+0.1}{s(s+1)}$ and $C(s) = \frac{20}{s}$ in response to a disturbance: (a) response to the unit step disturbance; (b) response to the unit ramp disturbance.

15.6   Exercises

1. Given $G(s) = \frac{20}{(s+2)(s+5)}$, determine a controller $C(s)$ such that $y_{ss} = 0$ for a unit step disturbance, and $|e_{ss}| \leq 0.1$ for a unit ramp input.

2. Given $G(s) = \frac{0.5s+1}{s^2+3s+20}$, determine a controller $C(s)$ such that $|e_{ss}| = 0$ for a unit ramp input, and $|e_{ss}| \leq 0.1$ for a unit parabolic input.

3. Given $G(s) = \frac{s+2}{s^2+5s+1}$, determine a controller $C(s)$ such that $y_{ss} = 0$ for a unit step disturbance, and $|e_{ss}| \leq 0.01$ for a unit ramp input.

4. Given $G(s) = \frac{s+2}{s^2+s+1}$, determine a controller $C(s)$ such that $y_{ss} = 0$ for a unit step disturbance, and $|e_{ss}| \leq 0.01$ for a unit ramp input.

5. Given $G(s) = \frac{2-s}{s^2+3s+5}$, determine a controller $C(s)$ such that $|e_{ss}| = 0$ for a unit step input, and $|e_{ss}| \leq 1$ for a unit ramp input.

16

Lecture 16—Controller design: transient performance

In this lecture, the design of the controller based on frequency compensation is dealt with. First, the two parameters fundamental to the design, namely the critical frequency and the phase margin, are discussed. Then, the lead and lag compensator are introduced. The guidelines for the design of these compensators to meet the requirements in terms of critical frequency and phase margin are illustrated through several examples and exercises. As it will be discussed, the procedure is based on trial and error, and, is, in fact, often referred to as the trial and error method for compensator design in the frequency domain.

16.1 The critical frequency and the phase margin

Let us consider an open-loop transfer function $G(s)$ that satisfies the Bode criterion, namely it has no poles with positive real part and crosses the 0dB-axis only once, as in Fig. 16.1. In addition, let $F(s)$ be the complementary sensitivity, and let ω_c be the crossover or critical frequency, namely the frequency at which $|C(j\omega_c)G(j\omega_c)| = 1$.

Since, for $\omega < \omega_c$, $|1 + C(j\omega)G(j\omega)| \simeq |C(j\omega)G(j\omega)|$ and, for $\omega > \omega_c$, $|1 + C(j\omega)G(j\omega)| \simeq 1$, the complementary sensitivity $F(s)$ can be approximated as follows:

$$|F(j\omega)| = \frac{|C(j\omega)G(j\omega)|}{|1 + C(j\omega)G(j\omega)|} = \begin{cases} 1 & \omega \leq \omega_c \\ |C(j\omega)G(j\omega)| & \omega > \omega_c \end{cases} \tag{16.1}$$

The approximation is also illustrated in Fig. 16.1, which shows that $F(s)$ behaves as a low-pass filter with unit gain. The critical frequency ω_c well represents the bandwidth of the filter.

Assuming the closed-loop system is asymptotically stable (we will discuss later how to verify and ensure stability), the critical frequency ω_c serves as a key indicator of system performance. Recalling that the complementary sensitivity is the transfer function from $r(t)$ to $y(t)$, we conclude that the

DOI: 10.1201/9781003487289-16

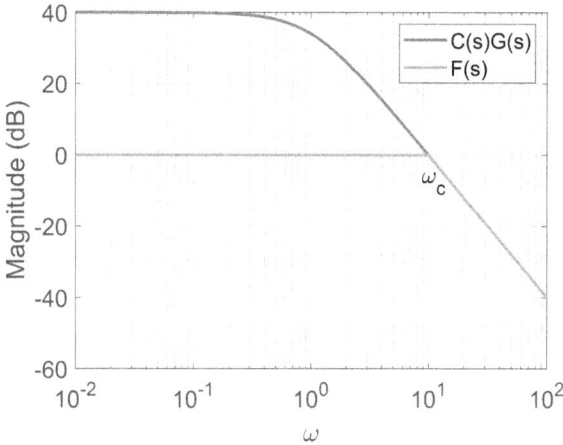

FIGURE 16.1
Magnitude Bode diagram of the open-loop function $C(s)G(s)$ and approximation (16.1) of the complementary sensitivity.

system will be able to track frequencies up to ω_c. This consideration would suggest using high critical frequencies. However, since the transfer function from $n(t)$ to $y(t)$ is $-F(s)$, it is clear that ω_c must be selected as a trade-off between the requirement to track the reference $r(t)$ and the need to reject the disturbance $n(t)$.

Another important consideration about ω_c is that, as it clearly appears from Fig. 16.1, it represents the natural frequency of the dominant poles of the closed-loop performance, and as such it impacts the dynamical performance, e.g., of the transient of the step input response of the system.

To understand whether the dominant poles are real or complex, or which is the damping coefficient associated with them, we need to focus on the behavior of $|F(j\omega)|$ at the critical frequency. In fact, this is the point around which the approximation introduces the largest error. Specifically, if the damping coefficient is small, then the magnitude of the complementary sensitivity will display a resonant peak. Let us assume that the closed-loop dominant poles are indeed complex, and let φ_c indicate the phase of $C(j\omega)G(j\omega)$ at ω_c. Then, $C(j\omega_c)G(j\omega_c) = |C(j\omega_c)G(j\omega_c)|e^{j\varphi_c} = e^{j\varphi_c}$ (since by definition of ω_c we have that $|C(j\omega_c)G(j\omega_c)| = 1$). Using this result in the expression of the complementary sensitivity, we get

$$\begin{aligned}|F(j\omega_c)| &= \frac{|C(j\omega_c)G(j\omega_c)|}{|1+C(j\omega_c)G(j\omega_c)|} = \frac{1}{|1+e^{j\varphi_c}|}\\ &= \frac{1}{\sqrt{(1+\cos\varphi_c)^2+\sin^2\varphi_c}} = \frac{1}{\sqrt{2(1+\cos\varphi_c)}}\end{aligned} \tag{16.2}$$

Since $m_\varphi = 180° - |\varphi_c|$, then

$$|F(j\omega_c)| = \frac{1}{\sqrt{2(1 - \cos m_\varphi)}} = \frac{1}{2 \sin \left(\frac{m_\varphi}{2}\right)} \tag{16.3}$$

By comparing this expression with the one found in Lecture 11, namely $|F(j\omega_c)| = 1/(2\xi)$ (rigorous for a second-order system without zeros and with unit gain, $\mu = 1$), we derive that

$$\xi = \sin \left(\frac{m_\varphi}{2}\right) \tag{16.4}$$

A common approximation of the previous expression is the following:

$$\xi \simeq \frac{m_\varphi}{2} \frac{\pi}{180°} \simeq \frac{m_\varphi}{100°} \tag{16.5}$$

that is derived by taking into account that m_φ is usually expressed in degrees.

This expression turns out to provide a good approximation of the relationship between ξ and m_φ up to $m_\varphi \simeq 75°$.

Altogether, ω_c and m_φ provide fundamental information on the closed-loop system performance, including the primary condition of closed-loop stability. According to the Bode criterion, on which this analysis is based, stability is guaranteed as long as $m_\varphi > 0$.

Similarly to what is done for the phase margin, also the critical frequency can be related to the parameters characterizing the two dominant complex and conjugate poles, namely the frequency ω_n and the damping coefficient ξ. To this aim, observe that the closed-loop function $F(s) = \frac{\omega_n^2}{s^2 + 2\xi\omega_n s + \omega_n^2}$ rigorously correspond to an open-loop system $C(s)G(s)$ given by

$$C(s)G(s) = \frac{\omega_n^2}{s(s + 2\xi\omega_n)} \tag{16.6}$$

In fact, we have that:

$$F(s) = \frac{C(s)G(s)}{1 + C(s)G(s)} = \frac{\frac{\omega_n^2}{s(s+2\xi\omega_n)}}{1 + \frac{\omega_n^2}{s(s+2\xi\omega_n)}} = \frac{\omega_n^2}{s^2 + 2\xi\omega_n s + \omega_n^2} \tag{16.7}$$

The critical frequency can be calculated from the condition $|C(j\omega_c) G(j\omega_c)| = 1$, namely

$$\frac{\omega_n^2}{\omega_c \sqrt{\omega_c^2 + 4\xi^2\omega_n^2}} = 1 \tag{16.8}$$

that yields

$$\omega_c = \omega_n \sqrt{-2\xi^2 + \sqrt{4\xi^4 + 1}} \tag{16.9}$$

16.2 Lead and lag compensators

Lead and lag compensators are elementary compensation functions formed by a pole and a zero. Depending on the relative position of the pole and the zero we distinguish between the *lead compensator*

$$C_{\text{lead}}(s) = \frac{1 + Ts}{1 + \frac{T}{\alpha}s} \tag{16.10}$$

and the *lag compensator*

$$C_{\text{lag}}(s) = \frac{1 + \frac{T}{\alpha}s}{1 + Ts} \tag{16.11}$$

where, in both cases, $T > 0$ and $\alpha > 1$.

The lead compensator has a zero in $z = -1/T$ and a pole in $p = -\alpha/T$. Since $\alpha > 1$, the lead compensator is also called zero-pole compensator. The Bode diagram of $C_{\text{lead}}(s)$ for different values of α is illustrated in Fig. 16.2. The key feature is the phase lead introduced by the compensator in the frequency range between the zero and the pole; this lead is beneficial for stability and increases as the parameter α becomes larger. The maximum lead occurs at a frequency $\omega_{\max}$ that is the geometric mean of the zero and the pole, i.e. $\omega_{\max} = \frac{1}{T\sqrt{\alpha}}$. At this frequency, the phase is given by:

$$\angle C_{\text{lead}}(j\omega_{\max}) = \arcsin \frac{\alpha - 1}{\alpha + 1} \tag{16.12}$$

The lag compensator has a pole in $p = -1/T$ and a zero in $z = -\alpha/T$. Since $\alpha > 1$, the lag compensator is also called pole-zero compensator. The Bode diagram of $C_{\text{lag}}(s)$ for different values of α is illustrated in Fig. 16.1. Notice that, since the lead and lag compensators only differ in the relative locations of their zeros and poles, the Bode diagram of Fig. 16.1 can be obtained by changing the sign of the diagram of Fig. 16.2. The lag compensator acts as a low-pass filter, with corner frequency given by the pole. The main effect is to introduce at selected frequencies an attenuation, whose intensity can be tuned by changing the parameter α, while T allows to tune the frequency range. The phase diagram shows that the compensator introduces a phase lag. The minimum occurs at $\omega_{\min} = \frac{1}{T\sqrt{\alpha}}$ and takes the value: $\angle C_{\text{lag}}(j\omega_{\min}) = -\arcsin \frac{\alpha-1}{\alpha+1}$.

Lead and lag compensators provide flexible tools to shape the frequency response of a system, and thus to control its performance. They can also be used in multiples, or in combination, e.g., the lead-lag compensator with transfer function:

$$C_{\text{lead-lag}}(s) = \frac{1 + T_{\text{lead}}s}{1 + \frac{T_{lead}}{\alpha_{lead}}s}\frac{1 + \frac{T_{\text{lag}}}{\alpha_{\text{lag}}}s}{1 + T_{\text{lag}}s} \tag{16.13}$$

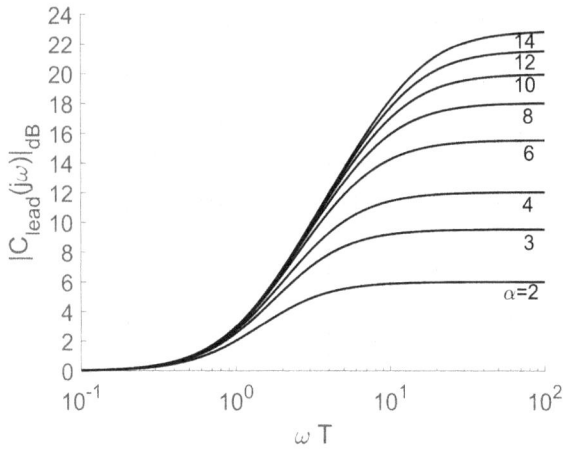

(a)

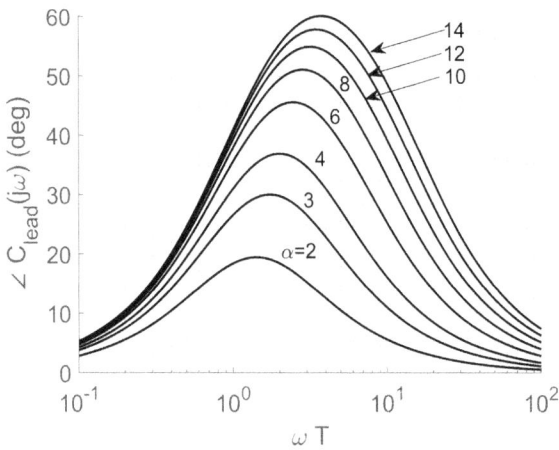

(b)

FIGURE 16.2
Bode diagram of the lead compensator $C_{\text{lead}}(s) = \frac{1+Ts}{1+\frac{T}{\alpha}s}$ for different values of α: (a) magnitude; (b) phase.

The impact of lead and lag compensators on closed-loop stability can be understood by analyzing their influence on the phase margin. Compensator design using frequency response operates under the assumption that the system to be controlled meets the hypotheses of the Bode criterion for stability, where a positive phase margin guarantees closed-loop stability. A lead compensator, introducing a zero at a frequency lower than its associated pole, enhances stability by increasing the phase margin, thereby improving the stability margin. On the other hand, a lag compensator introduces a pole at a frequency lower than its zero, which decreases the phase in the frequency range before the pole influence. This reduction in phase can negatively affect stability, making the lag compensator potentially detrimental to stability in certain cases.

The use of lead and lag compensators will be illustrated in the next section.

16.3   Frequency compensation design

We will study four scenarios:

- Case 1. Increasing m_φ without changing ω_c;

- Case 2. Increasing m_φ, under the assumption that ω_c can decrease;

- Case 3. Increasing ω_c;

- Case 4. Decreasing ω_c.

The analysis of these four scenarios is not intended to be exhaustive of all possible situations but rather to provide guidelines on the use of lead and lag compensators and what can be achieved with them.

16.3.1   Case 1

In this case, the purpose of the compensation is to increase m_φ without changing ω_c. This can be done by using a lead compensator where T is selected such that $\omega_c T = 1$. In this way, the introduction of the lead compensator yields a very small (ideally zero) change in the magnitude of the open-loop transfer function around ω_c (see Fig. 16.2(a)). At the same time, in correspondence of this frequency, the phase of the lead compensator is already significantly larger than zero. This phase difference can be used to achieve the desired phase margin.

The procedure is illustrated in the following example.

Example 16.1 ─────────────────────────────────

Consider the system $G(s) = \frac{20}{s(1+0.5s)}$ and assume that the specifications are $m_\varphi = 45°$ and $\omega_c \simeq 6$ rad/s.

First, the Bode diagram of the open-loop system without control, namely $G(s)$, is considered, in order to evaluate the values of ω_c and m_φ in the absence of control. The Bode diagram is illustrated in Fig. 16.4 where the curve without compensation shows $\omega_c \simeq 6$ rad/s with $m_\varphi = 18°$. Thus, the critical frequency is acceptable, but the phase margin is not. To increase the phase margin, a lead compensator can be used. The selection of the design parameters of this compensator is carried out by considering the following factors. As we do not want to introduce large changes in the magnitude of the open-loop transfer function around the frequency $\omega_c \simeq 6$ rad/s, we select T such that $\omega_c T = 1$. In this way, since $|C_{\text{lead}}(j\omega)| \simeq 4 \div 5\text{dB}$ as shown in Fig. 16.2(a), the magnitude of the open-loop transfer function will not significantly change after the introduction of the compensator. Then, the parameter α is chosen considering that without compensation $m_\varphi = 18°$, so a lead compensator introducing a phase difference of at least $30°$ is needed. From Fig.. 16.2(b) we find that at $\omega T = 1$ the compensator with $\alpha = 4$ introduce the desired phase difference. Summing up, a lead compensator with $T = 1/6$ and $\alpha = 4$ is selected, namely $C_{\text{lead}} = \frac{1+s/6}{1+2s/3}$. The use of this compensator yields the Bode diagram of Fig. 16.4 (curve with compensation), for which $\omega_c \simeq 7.7$ rad/s with $m_\varphi = 48.8°$. After compensation the frequency ω_c remains almost unchanged, while the phase margin reaches the desired value.

16.3.2 Case 2

In this case, the purpose of the compensation is to increase m_φ up to the desired value $m_{\varphi,des}$, being possible to change the value of ω_c. To this aim, a lag compensator can be used. The idea is to decrease the value of ω_c by introducing an attenuation in the magnitude of the open-loop frequency response, such that the compensated magnitude Bode diagram crosses the x-axis at a frequency $\omega_{t,n}$ for which the phase is $\angle C(j\omega_{t,n})G(j\omega_{t,n}) = m_{\varphi,des} - m_\varphi$. This can be achieved by selecting T in the lag compensator such that $\omega_{t,n}T = 20 \div 100$, which corresponds to a frequency range where the phase shift introduced is minimal, but the change in magnitude is significant.

The procedure is illustrated in the following example.

Example 16.2 _____

Consider again the system $G(s) = \frac{20}{s(1+0.5s)}$ analyzed in Example 16.1 and assume that the control goal is to have $m_\varphi = 45°$.

We have already seen that, without control, $\omega_c \simeq 6$ rad/s and $m_\varphi = 18°$. To obtain the desired phase margin, it is possible to move ω_c such that Bode diagram of the magnitude crosses the x-axis in correspondence with a frequency for which the phase is greater than $-135°$, thus guaranteeing the desired phase margin. To do this, we introduce a lag compensator and select its parameters taking into account the following considerations. From the Bode diagram of $G(s)$ (Fig. 16.4, curve without compensation) we find that $\angle G(j\omega) = -135°$ at $\omega = 2$ rad/s. However, we also consider that the lag compensator will introduce a negative phase difference that deteriorates the phase margin (of about $5° \div 10°$, as shown in Fig. 16.3(a)). Thus, we select a slightly smaller frequency, namely $\omega = 1.5$ rad/s where $\angle G(j\omega) \simeq -125°$. Next, one has to calculate the magnitude at this frequency (either graphically from the Bode diagram or directly from $G(s)$). We get: $|G(j1.5)|_{dB} \simeq 21\text{dB}$. At this point we can select T such that $\omega_{t,n}T = 50$ (where $\omega_{t,n} = 1.5$ rad/s) and α such that at $\omega_{t,n}T = 50$ the lag compensator introduces the desired magnitude difference, namely -21dB. As shown in Fig. 16.3(a), this can be obtained by selecting $\alpha = 12$. Summing up, a lag compensator with $T = 50/1.5$ and $\alpha = 12$ is selected, namely $C_{\text{lag}} = \frac{1+50s/18}{1+50s/1.5}$. The use of this compensator yields the Bode diagram of Fig. 16.5 (curve with compensation), for which $\omega_c \simeq 1.4$ rad/s with $m_\varphi = 42°$.

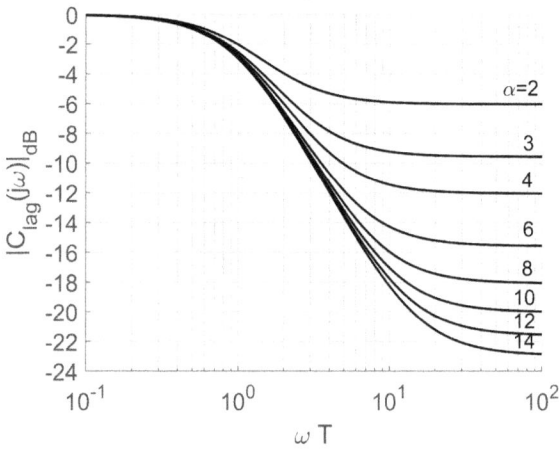

(a)

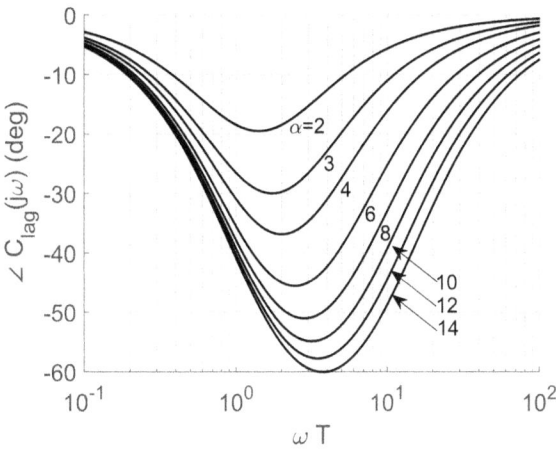

(b)

FIGURE 16.3
Bode diagram of the lag compensator $C_{\text{lag}}(s) = \frac{1+\frac{T}{\alpha}s}{1+Ts}$ for different values of α: (a) magnitude; (b) phase.

16.3.3 Case 3

In this scenario, the purpose of the controller is to increase the critical frequency. Let ω_c indicate the desired value for the critical frequency. The first

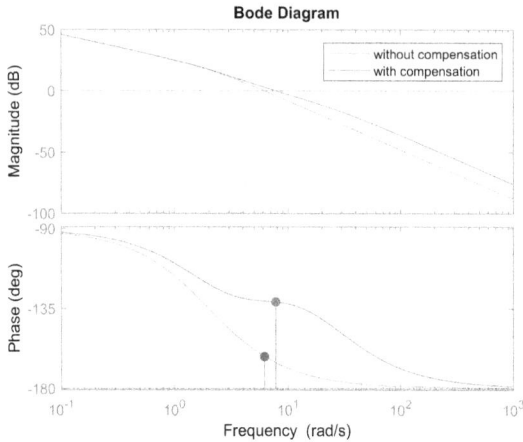

FIGURE 16.4
Frequency compensation design: Case 1. Bode diagram without and with compensation for Example 16.1.

step to perform is to evaluate the phase at the desired critical frequency. Two scenarios may occur. In the first case, $\angle G(j\omega_c) > m_\varphi - 180°$, which means that the only action to perform is to let the Bode diagram cross the

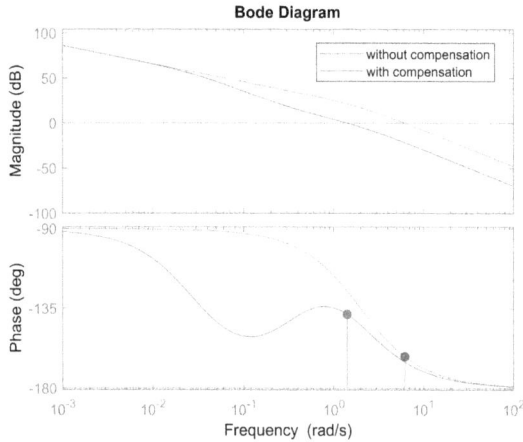

FIGURE 16.5
Frequency compensation design: Case 2. Bode diagram without and with compensation for Example 16.2.

0dB axis at ω_c. Once done this, the phase margin is already greater or equal to the desired one, and no other corrective actions need to be carried out. Therefore, increasing the controller gain K_c is sufficient as this will move toward right the critical frequency, up to the desired value. Otherwise, in the second scenario, $\angle C(j\omega_c)G(j\omega_c) < m_\varphi - 180°$. In this case, we can use a lead compensator that introduces the needed phase difference at ω_c, that is, $\angle C(j\omega_c) = m_\varphi - 180° - \angle G(j\omega_c)$. The scenario is similar to case 1, but with the difference that the increased magnitude, associated with the action of the lead compensator, can be used for our purpose of increasing the critical frequency. If this is not enough to compensate for $|G(j\omega_c)|$, then we can also introduce a gain factor equal to: $K_{c,dB} = -|G(j\omega_c)|_{dB} - |C(j\omega_c)|_{dB}$. Note that in both cases increasing the controller gain does not contrast with the requirements on the steady-state performance.

The procedure is illustrated in the following example.

Example 16.3 _____

Consider again the system $G(s) = \frac{20}{s(1+0.5s)}$ and assume that the specifics are $m_\varphi = 45°$ and $\omega_c \simeq 18$ rad/s.

This time we need to increase the critical frequency as in the absence of control the magnitude Bode diagram crosses the 0 dB axis at 6.2 rad/s. First, we evaluate the phase at the desired critical frequency. We find $\angle G(j18) = -175°$. Since the desired phase margin is $m_\varphi = 45°$, we need to introduce a lead compensator. At the same time, in order to have the open-loop magnitude Bode diagram cross the 0 dB axis at 18 rad/s, we have to increase the magnitude either for the whole diagram (increasing the gain), or locally (around the desired critical frequency with the action of a lead compensator). Since $|G(j18)|_{dB} = -18$dB, we have to compensate for this quantity.

We proceed as follows. We consider the maximum phase difference that can be introduced by a lead compensator as in Eq. (16.12). Setting $\alpha = 6$ yields a maximum phase equal to 45.6° (see also Fig. 16.2(b)). The maximum occurs at $\omega_c = \sqrt{\alpha}/T$; hence we select $T = \sqrt{6}/18$. Correspondingly, as shown in Fig. 16.2(a) the lead compensator will introduce at this frequency an increase of the magnitude equal to 8dB. The remaining 10dB are compensated through the controller gain which is set to $K_c = 10^{10/20} = 3.16$. Summing up, we select the following controller $C(s) = 3.16\frac{1+0.14s}{1+0.023s}$ and obtain the Bode diagram of Fig. 16.6 (curve with compensation), for which $\omega_c \simeq 17.5$ rad/s with $m_\varphi = 52°$.

16.3.4 Case 4

The last scenario we consider is when the goal is to decrease the value of ω_c. If there are no constraints on the static gain of the open-loop transfer function, deriving from the steady-state performance, then the value of ω_c can be reduced by decreasing the static gain. However, in the more general situation, this requires the use of a lag compensator as for case 2. At this point, we should evaluate the phase at the desired critical frequency, and as for case 3 determine whether a lead compensator is also required or not. Summing up, in the more general situation, case 4 will require the use of a lead-lag compensator, whose parameter can be selected following the guidelines described in the previous cases.

The procedure is illustrated in the following example.

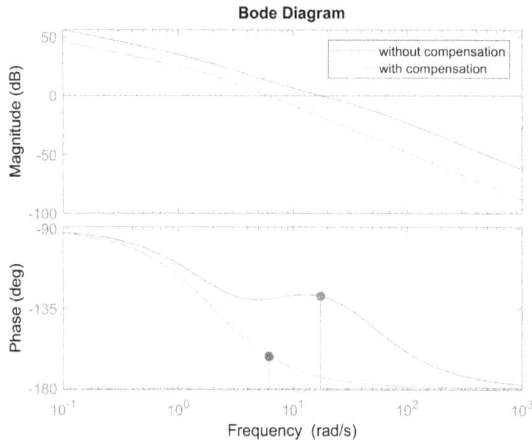

FIGURE 16.6

Frequency compensation design: Case 3. Bode diagram without and with compensation for Example 16.3.

Example 16.4 _____

Consider the system $G(s) = \frac{40(1+0.2s)}{s^2(1+0.05s)}$ and assume that the specifications are $m_\varphi = 40°$ and $\omega_c \simeq 3$ rad/s. In addition, assume also that the open-loop static gain cannot be decreased.

First, we analyze the Bode diagram of the open-loop system without control, namely $G(s)$. The diagram is illustrated in Fig. 16.4 where the curve without compensation shows that $\omega_c \simeq 8.5$ rad/s with $m_\varphi = 36.5°$. Thus, the critical frequency needs to be decreased. Next, we calculate $|G(j\omega)|_{\mathrm{dB}}$ and $\angle G(j\omega)$ at the desired critical frequency, that is, for $\omega = \omega_c = 3$ rad/s. We find: $|G(j\omega)|_{\mathrm{dB}} = 13.7$dB and $\angle G(j\omega) = -157°$. We need to use a lead-lag compensator. To select the parameters of this compensator, we proceed as for cases 1 and 2. In the selection of the parameters of the lead compensator, we will take into account that the lag compensator will also introduce a negative phase difference, and, similarly, in the selection of the parameters of the lag compensator we will take into account that the lead compensator will also introduce an increase in the magnitude of the open-loop frequency response. More specifically, we select a lead compensator able to compensate for a phase difference of about 25° at $\omega_c T = 1$, i.e., $\alpha_{\text{lead}} = 4$, and a lag compensator able to decrease the magnitude of about $-18dB$ at $\omega_c T = 50$, i.e., $\alpha_{\text{lead}} = 8$. The resulting controller is given by: $C(s) = \frac{(1+s/3)(1+50s/24)}{(1+s/12)(1+50s/3)}$. The use of this controller yields the Bode diagram in Fig. 16.7 (curve with compensation), for which $\omega_c \simeq 2.8$ rad/s with $m_\varphi = 42°$.

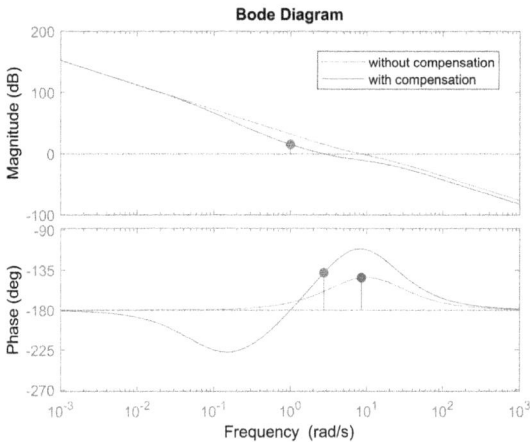

FIGURE 16.7
Frequency compensation design: case 4. Bode diagram without and with compensation for Example 16.4.

16.4   MATLAB examples

We illustrate here the MATLAB® commands for the four examples discussed in the lecture. All of them use commands that we have already illustrated, namely **bode** to draw the Bode diagram and **margin** to calculate the stability margins.

MATLAB® exercise 16.1 _____

For Example 16.1, the following commands are used:

```
s=tf('s')
G=20/(s*(1+0.5*s));
[Gm,pm,wpi,wt]=margin(G)
T=1/6;
alpha=4;
Clead=(1+T*s)/(1+T*s/alpha);
[Gm,pm,wpi,wt]=margin(Clead*G)
figure,bode(G,':')
hold on
bode(Clead*G)
legend('without compensation','with compensation')
```

MATLAB® exercise 16.2

For Example 16.2, the following commands are used:

```
s=tf('s')
G=20/(s*(1+0.5*s));
[Gm,pm,wpi,wt]=margin(G)
T=50/1.5;
alpha=12;
Clag=(1+T*s/alpha)/(1+T*s);
[Gm,pm,wpi,wt]=margin(Clag*G)
figure,bode(G,':')
hold on
bode(Clag*G)
legend('without compensation','with compensation')
```

MATLAB® exercise 16.3

For Example 16.3, the following commands are used:

```
s=tf('s')
G=20/(s*(1+0.5*s));
[Gm,pm,wpi,wt]=margin(G)
T=2.5/18;
alpha=6;
gainC=3.16;
Clead=gainC*(1+T*s)/(1+T*s/alpha);
[Gm,pm,wpi,wt]=margin(Clead*G)
figure,bode(G,':')
hold on
bode(Clead*G)
legend('without compensation','with compensation')
```

MATLAB® exercise 16.4

For Example 16.4, the following commands are used:

```
s=tf('s')
G=40*(1+0.2*s)/(s^2*(1+0.05*s));
[Gm,pm,wpi,wt]=margin(G)
T=1/3;
alpha=4;
Clead=(1+T*s)/(1+T*s/alpha);
Tlag=50/3;
alphalag=8;
Clag=(1+Tlag*s/alphalag)/(1+Tlag*s);
[Gm,pm,wpi,wt]=margin(Clead*Clag*G)
figure,bode(G,':')
hold on
bode(Clead*Clag*G)
legend('without compensation','with compensation')
```

16.5   Exercises

1. Given $G(s) = \frac{10}{s(s+5)}$, design a controller $C(s)$ such that $m_\varphi = 50°$ and $\omega_c = 20$ rad/s.

2. Given $G(s) = \frac{80}{s(s+1)(s+8)}$, design a controller $C(s)$ such that $m_\varphi = 45°$.

3. Given $G(s) = \frac{50(s+1)}{s(s^2+4s+25)}$, design a controller $C(s)$ such that $m_\varphi = 50°$ and $\omega_c = 20$ rad/s.

4. Given $G(s) = \frac{60(50s+1)}{s^2(s+10)}$, design a controller $C(s)$ such that $m_\varphi = 60°$ and $\omega_c = 20$ rad/s.

5. Given $G(s) = \frac{20}{s(3s+3)(0.1s+1)}$, design a controller $C(s)$ such that $m_\varphi = 45°$ and $\omega_c \simeq 3$ rad/s.

17

Lecture 17—Controller design: complete procedure and examples

This lecture deals with the complete design of the controller. The goal of the design is to obtain a controller satisfying a series of performance specifications, usually given in terms of characteristics of the closed-loop step response, e.g., the steady-state error to given canonical signals, the overshoot, the rise time, and so on. After briefly introducing the procedure, several worked examples of the design of the compensator are illustrated.

17.1 Design of the compensator

The procedure for the design of the compensator discussed in this lecture is an iterative one. It is based on the frequency compensation method described in the two previous lectures. It starts with the analysis of the performance specifications that, for instance, can be given in terms of desired characteristics of the closed-loop step response. These specifications need to be transformed into the corresponding specifications for the open-loop transfer functions, namely the system type, the open-loop gain, the critical frequency, and the phase margin. At this point, first the compensator to satisfy the steady-state performance specifications is derived as described in Lecture 15. Then, the compensator to satisfy the transient performance specifications is designed by following the guidelines discussed in Lecture 16. The next step is to apply the obtained compensator to the controlled system and evaluate the performance specifications given as the goal of the control, e.g., by calculating the closed-loop step response. The evaluation of these performance specifications provides indications if the design of the compensator needs another iteration or should be stopped. The analysis of stability leverages the Bode criterion, such that, even if a target value for m_φ is not given, in any case in applying the procedure it should be guaranteed that $m_\varphi > 0$. This also means that the procedure can be applied only if the assumptions of the Bode criterion are satisfied.

DOI: 10.1201/9781003487289-17

17.2   Performance specifications

Here, we summarize the relationships existing from the different performance specifications that can be given for a system.

We begin with steady-state performance specifications. These are illustrated in Lecture 15 and, ultimately, determine the system type and the gain of the open-loop transfer function $C(s)G(s)$:

steady-state step/ramp/parabolic input error $\Rightarrow$ system type and gain (of the open-loop transfer function)

Next, we consider the transient performance specifications. These can be given in terms of the step response of the closed-loop system, in terms of the closed-loop frequency response, or in terms of the open-loop frequency response. In all cases the reference model is a second-order system with complex and conjugate poles, with unit gain and without any zero, namely $F(s) = \frac{\omega_n^2}{s^2+2\xi\omega_n s+\omega_n^2}$.

- Closed-loop step response.

  Rise time t_r (see Eq. (10.23), Lecture 10):

  $$t_r \simeq \frac{1.8}{\omega_n} \tag{17.1}$$

  Settling time $t_{s,1\%}$ (see Eq. (10.26), Lecture 10):

  $$t_{s,1\%} \simeq \frac{4.6}{\xi\omega_n} \tag{17.2}$$

  Overshoot $s_\%$ (see Eq. (10.25) and Fig. 10.7, Lecture 10):

  $$s_\% = 100e^{-\frac{\pi\xi}{\sqrt{1-\xi^2}}} \tag{17.3}$$

  These equations express the relationships between performance specifications of the closed-loop step response and the positions of the dominant poles in the complex plane (see also Fig. 10.4, Lecture 10).

- Closed-loop frequency response.

  Magnitude peak M_r at the resonance frequency $\omega_r = \omega_n\sqrt{1-2\xi^2}$ (see Eq. (11.29), Lecture 11):

  $$M_r = |F(j\omega_r)|_{dB} = \frac{1}{2|\xi|\sqrt{1-\xi^2}} \tag{17.4}$$

Bandwidth B_3 (see Eq. (11.31), Lecture 11):

$$B_3 = \omega_n \sqrt{(1 - 2\xi^2) + \sqrt{4\xi^4 - 4\xi^2 + 2}} \tag{17.5}$$

Eqs. (17.4) and (17.5) express a relationship between specifications in the frequency domain and positions of the poles, for the closed-loop system. It is also possible to link M_r and B_3 with the specifications in the time domain through some approximated relationships:

$$\frac{1 + s_\%/100}{M_r} \simeq 0.85 \tag{17.6}$$

in a wide range of values of ξ, and

$$B_3 t_s \simeq 3 \tag{17.7}$$

- Open-loop frequency response.

  Critical frequency ω_c (see Eq. (16.9), Lecture 16):

$$\omega_c = \omega_n \sqrt{-2\xi^2 + \sqrt{4\xi^4 + 1}} \tag{17.8}$$

Phase margin m_φ (see Eq. (16.5), Lecture 16):

$$m_\varphi = 100\xi \tag{17.9}$$

Eqs. (17.8) and (17.9) relate the open-loop frequency response specifications to the characteristic parameters of the dominant poles, ξ and ω_n.

Altogether, the relationships discussed in this section allow for translating specifications given for the closed-loop frequency response or step response into constraints on ω_c and m_φ. These two parameters serve as the starting point for applying the procedure outlined in Lecture 16 to achieve the desired performance specifications.

17.3   Procedure for the frequency response-based controller design

The procedure for the frequency response-based controller design consists of a series of steps that are summarized as follows:

1. Design the compensator to satisfy the steady-state performance, following the guidelines of Lecture 15.

2. From the given specifications derive the conditions on m_φ and ω_c, using the relationships summarized in Sec. 17.2.

3. Draw the Bode diagram of the open-loop system that includes the compensator designed for the steady-state performance.

4. Use lead, lag, or lead-lag compensators to meet the requirements on m_φ and ω_c, following the guidelines of Lecture 16.

5. Check if the dynamic specifications are met or not, by calculating the closed-loop response, the step response, and the frequency response. If the behavior is satisfactory, the design procedures is finished. Otherwise, readjust the conditions on m_φ and ω_t, and go back to step 3.

As already discussed, the procedure can be applied to systems that satisfy the assumptions of the Bode criterion. The resulting controller is characterized by a transfer function that includes a gain, a number of poles at the origin, and a series of lead, lag, and/or lead-lag compensators. Note that achieving the requirements for steady-state performance often necessitates increasing the number of poles at the origin of the open-loop transfer function. This generally has a detrimental effect on closed-loop stability and consequently requires greater effort in designing the compensator to meet transient specifications and ensure stability.

17.4 Examples of compensator design

In this section, we will illustrate several examples of the design of the compensator using the procedure described in the previous sections.

MATLAB® exercise 17.1 _____

Given the system $G(s) = \dfrac{\frac{s}{2}+1}{\left(\frac{s}{3}+1\right)^2}$ design a controller that ensures the following performance specifications:

- null steady-state error of the step response;
- steady-state error to the unit ramp, $e_{ss} \le 0.1$;
- percentage overshoot $s_\% \le 20\%$;
- bandwidth $B_3 \simeq 3$ rad/s.

As a first step, we translate the given performance specifications into conditions on the critical frequency ω_c and the phase margin m_φ. The relationship between the percentage overshoot and ξ is given in Eq. (17.3) and is illustrated in Fig. 10.7. From the curve in Fig. 10.7 we derive that $\xi \ge 0.45$. Let us set $\xi = 0.45$. Next, we use Eq. (17.5) to derive ω_n, given $B_3 \simeq 3$ rad/s:

$$\omega_n = B_3 / \sqrt{(1 - 2\xi^2) + \sqrt{4\xi^4 - 4\xi^2 + 2}} = 2.26 \text{ rad/s} \qquad (17.10)$$

We replace this value, along with $\xi = 0.45$, in Eq. (17.8) to obtain

$$\omega_c = \omega_n \sqrt{-2\xi^2 + \sqrt{4\xi^4 + 1}} = 1.86 \text{ rad/s} \qquad (17.11)$$

Finally, from Eq. (17.9) we get that $m_\varphi = 100\xi \geq 45°$.

As a second step, we design the controller to meet the steady-state performance specifications. Since $G(s)$ has no pole at the origin and the open-loop system needs to be at least type 1, $C_1(s)$ should have at least a pole at the origin. The requirement that the steady-state error to the unit ramp has to be $e_{ss} \leq 0.1$ yields that the controller gain has to be greater than or equal to 10 (notice that $G(0) = 1$). Hence, the compensator for the steady-state performance specifications is given by $C_1(s) = \frac{10}{s}$.

As a third step, we design the compensator to ensure the dynamical performance specifications, namely $\omega_c \simeq 1.86$ rad/s and $m_\varphi \geq 45°$. To this aim, we draw the Bode diagram of the transfer function $C_1(s)G(s) = \frac{10\left(\frac{s}{2}+1\right)}{s\left(\frac{s}{3}+1\right)^2}$. This is illustrated in Fig. 17.1, from which we get that $\omega_c \simeq 6.2$ rad/s and $m_\varphi \simeq 33.8°$. From the same curve, we also find that at the desired critical frequency, $\omega \simeq 1.85$ rad/s, the magnitude is 14.6dB and the phase is $-110°$. We can decrease the critical frequency by using a lag compensator with $\alpha = 6$ and $\omega_c T = 20$. This introduces a phase lag of about $15°$ that can be tolerated since the initial phase at $\omega \simeq 1.85$ rad/s is $-110°$. The Bode diagram of the system compensated with $C_1(s) = \frac{10}{s}$ and $C_{\text{lag}}(s) = \frac{10.81s+6}{64.86s+6}$ is also shown in Fig. 17.1. We find $\omega_c \simeq 1.73$ rad/s and $m_\varphi \simeq 56°$.

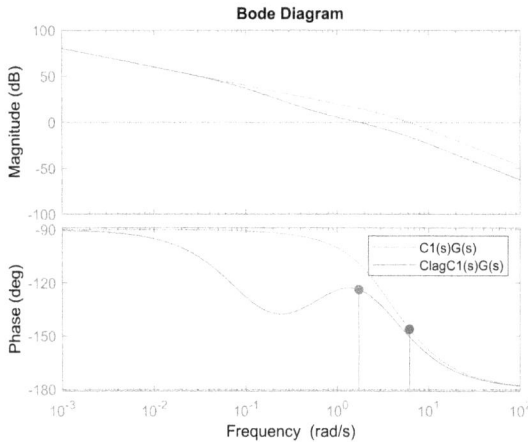

FIGURE 17.1
Bode diagram for the system with only the compensator for steady-state performance specifications, namely $C_1(s)G(s)$, and with the complete compensator, namely namely $C_1(s)C_{\text{lag}}(s)G(s)$, in Exercise 17.1.

Next, in order to check that the system meets the performance specifications on the percentage overshoot $s_\%$ and the bandwidth B_3, we calculate the step response and the Bode diagram of the closed-loop system, $F(s) = \frac{C_1(s)C_{\text{lag}}(s)G(s)}{1+C_1(s)C_{\text{lag}}(s)G(s)}$. They are illustrated in Fig. 17.2(a) and (b), respectively. From Fig. 17.2(a) we observe that $s_\% \leq 20\%$, while from Fig. 17.2(b) we find that $B_3 \simeq 2.9$ rad/s.

The results can be replicated using the following MATLAB® commands:

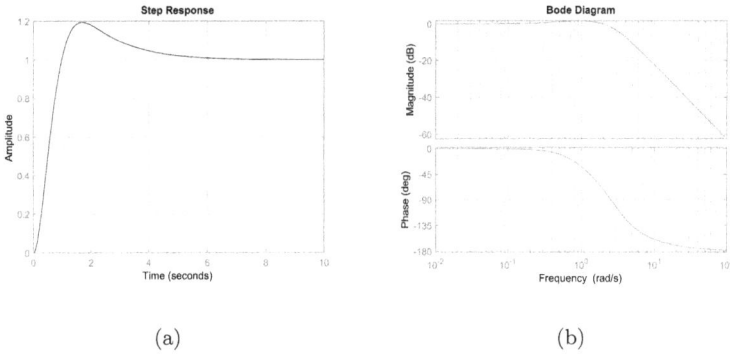

(a) (b)

FIGURE 17.2
Behavior of the closed-loop system $F(s) = \frac{C_1(s)C_{\text{lag}}(s)G(s)}{1+C_1(s)C_{\text{lag}}(s)G(s)}$ in Exercise 17.1:
(a) step response; (b) Bode diagram.

```
s=tf('s');
G=(s/2+1)/(s/3+1)^2;
C1=10/s;
figure,bode(C1*G,':')
T=20/1.85;
alpha=6;
Clag=(1+T*s/alpha)/(1+T*s);
hold on
bode(Clag*C1*G)
F=Clag*C1*G/(1+Clag*C1*G)
figure,step(F,'k')
w=logspace(-2,2,1001);
figure,bode(F,w)
```

MATLAB® exercise 17.2 _____

Given the system $G(s) = \frac{\frac{s}{10}+1}{s^2}$ design a controller that ensures the following performance specifications:

- null steady-state error for the response to the step and ramp input;
- steady-state error to the unit parabolic input, $e_{ss} \leq 0.1$;
- rise time $t_r \leq 50$ ms;
- maximum frequency peak $M_r \leq 1.2$.

We begin by deriving from the given performance specifications the conditions on the critical frequency ω_c and the phase margin m_φ. From Eq. (17.4) we derive that $M_r \leq 1.2$ requires $\xi \geq 0.48$. Let us fix $\xi = 0.48$. Next, using the specification on the rise time t_r, we apply Eq. (10.23) to derive that $\omega_n \geq 36$ rad/s. We replace this value, along with $\xi = 0.48$, in Eq. (17.8) to obtain

$$\omega_c = \omega_n \sqrt{-2\xi^2 + \sqrt{4\xi^4 + 1}} = 28.8 \text{ rad/s} \qquad (17.12)$$

Finally, from Eq. (17.9) we get that $m_\varphi = 100\xi \geq 48°$.
The second step consists of the design of the controller to meet the steady-state performance specifications. A type 2 system is required to obtain a null steady-state error

to the step and ramp input. Since $G(s)$ already has two poles at the origin, $C_1(s) = k_1$ may be used, that is, a static gain. To meet the specification on the parabolic input, $k_1 \geq 10$ is required. Let us select $k_1 = 10$ and proceed to the design of the compensator for the other performance specifications.

First, we draw the Bode diagram of the transfer function $C_1(s)G(s) = \frac{10(\frac{s}{10}+1)}{s^2}$. The result is shown in Fig. 17.3, from which we get that $\omega_c \simeq 3.2$ rad/s and $m_\varphi \simeq 18°$. From the diagram, we find that at the desired critical frequency, $\omega \simeq 29$ rad/s, the magnitude is -28.8dB and the phase is $-109°$. Increasing the gain of a factor equal to $10^{28.8/20} = 27.6$ would increase the critical frequency, obtaining at the same time a phase margin slightly larger than the desired one. Let $k = 27.6$, we now draw the Bode diagram of the transfer function $kC_1(s)G(s) = \frac{276(\frac{s}{10}+1)}{s^2}$ and check the specifications on ω_c and m_φ. The diagram is also shown in Fig. 17.1. We get $\omega_c \simeq 29.2$ rad/s and $m_\varphi \simeq 71°$.

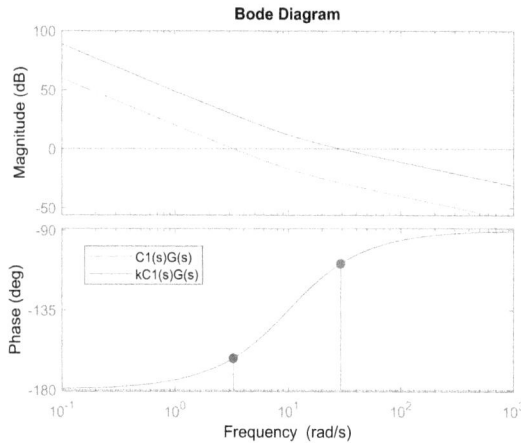

FIGURE 17.3

Bode diagram for the system with only the compensator for steady-state performance specifications, namely $C_1(s)G(s)$, and with the complete compensator, namely namely $kC_1(s)G(s)$, in Exercise 17.2.

To check that the system meets the performance specifications on the rise time t_r and the maximum frequency peak M_r, we calculate the step response and the Bode diagram of the closed-loop system, $F(s) = \frac{kC_1(s)G(s)}{1+kC_1(s)G(s)}$. They are illustrated in Fig. 17.4(a) and (b), respectively. We find $t_r = 47.8$ ms and $M_{r,dB} = 1.65$dB, that is, $M_r = 10^{1.65/20} = 1.21$.

The results can be replicated using the following MATLAB® commands:

```
s=tf('s');
G=(s/10+1)/s^2;
C1=10;
figure,bode(C1*G,':')
k=27.6
hold on
bode(k*C1*G)
legend('C1(s)G(s)','kC1(s)G(s)')
F=k*C1*G/(1+k*C1*G)
```

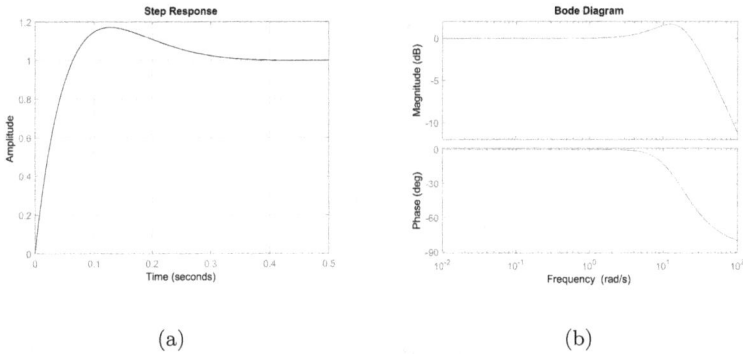

(a)   (b)

FIGURE 17.4
Behavior of the closed-loop system $F(s) = \frac{kC_1(s)G(s)}{1+kC_1(s)G(s)}$ in Exercise 17.2: (a) step response; (b) Bode diagram.

```
figure,step(F,'k')
stepinfo(F,'SettlingTimeThreshold',0.01)
w=logspace(-2,2,1001);
figure,bode(F,w)
```

17.5 Indirect design of a digital controller

The design of a digital controller, to be used in the control scheme of Fig. 14.2, can be performed by, first, deriving the compensator in the s-domain, and, then, obtaining the transfer function of the digital controller from the analog one. The second step is accomplished by choosing a suitable sampling time T_s, and applying the transformation $s = \frac{2}{T}\frac{z-1}{z+1}$. We illustrate the procedure through an exercise.

MATLAB® exercise 17.3 _____

Given the system $G(s) = \frac{s+2}{s^2+4s+3}$ design a digital controller that ensures the following performance specifications:

- null steady-state error of the step response;
- steady-state error to the ramp $|e_{ss}| \leq 1$;
- percentage overshoot $s_\% \leq 5\%$;
- bandwidth $B_3 \simeq 5$ rad/s.

Let us start by deriving the conditions on the critical frequency ω_c and the phase margin m_φ. Following the same steps of the previous exercise, we derive that $\xi = 0.7$. In addition, given $B_3 \simeq 5$ rad/s, we have that

$$\omega_n = B_3 / \sqrt{(1 - 2\xi^2) + \sqrt{4\xi^4 - 4\xi^2 + 2}} = 4.95 \text{ rad/s} \qquad (17.13)$$

that yields

$$\omega_c = \omega_n \sqrt{-2\xi^2 + \sqrt{4\xi^4 + 1}} = 3.21 \text{ rad/s} \qquad (17.14)$$

Finally, from Eq. (17.9) we get that $m_\varphi = 100\xi \geq 70°$.

Next, the controller to meet the steady-state performance specifications is designed. Since $G(s)$ has no pole at the origin and the open-loop system should be at least type 1, $C_1(s)$ should have a pole at the origin. The requirement on the steady-state error to the ramp input yields a controller gain $C(0) = 1/G(0) = 1.5$. Hence, the compensator for the steady-state performance specifications is given by $C_1(s) = \frac{1.5}{s}$.

To design the compensator for the dynamical performance specifications, namely $\omega_c \simeq 3.21$ rad/s and $m_\varphi \geq 70°$, we first draw the Bode diagram of the transfer function $C_1(s)G(s) = \frac{1.5(s+2)}{s(s^2+4s+3)}$. This is illustrated in Fig. 17.5, from which we get that $\omega_c \simeq 0.81$ rad/s and $m_\varphi \simeq 58°$. From the same curve, we also find that at the desired critical frequency, $\omega \simeq 3.21$ rad/sec, the magnitude is -18.4dB and the phase is $-152°$. We can increase the phase margin by using a lead compensator with $\alpha = 12$ and $\omega_c T = 1.2$. The Bode diagram of the system compensated with $C_1(s) = \frac{1.5}{s}$ and $C_{\text{lead}}(s) = \frac{4.49s+1}{0.37s+12}$ is also shown in Fig. 17.5. Finally a gain $K_1 = 5.6$ is included in order to obtain $\omega_c \simeq 3.33$ rad/s and $m_\varphi \simeq 73°$, as shown in Fig. 17.5.

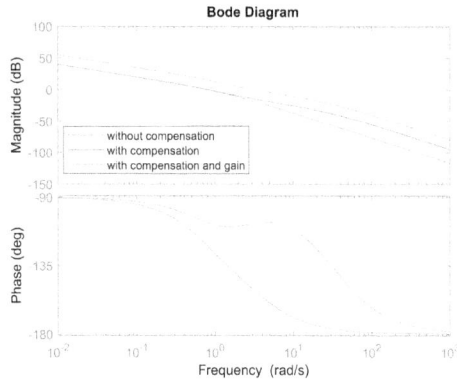

FIGURE 17.5

Bode diagram for the system with only the compensator for steady-state performance specifications, namely $C_1(s)G(s)$, with the compensator $C_1(s)C_{\text{lead}}(s)G(s)$, and the complete compensator $K_1C_1(s)C_{\text{lead}}(s)G(s)$ in Exercise 17.3.

In order to check that the system meets the performance specifications on the percentage overshoot $s_\%$ and the bandwidth B_3, we calculate the step response and the Bode diagram of the closed-loop system, $F(s) = \frac{K_1C_1(s)C_{\text{lead}}(s)G(s)}{1+K_1C_1(s)C_{\text{lead}}(s)G(s)}$. They are reported in Fig. 17.6(a) and (b), respectively, where we find an overshoot of about 7% and a bandwidth $B_3 \simeq 4.3$ rad/s.

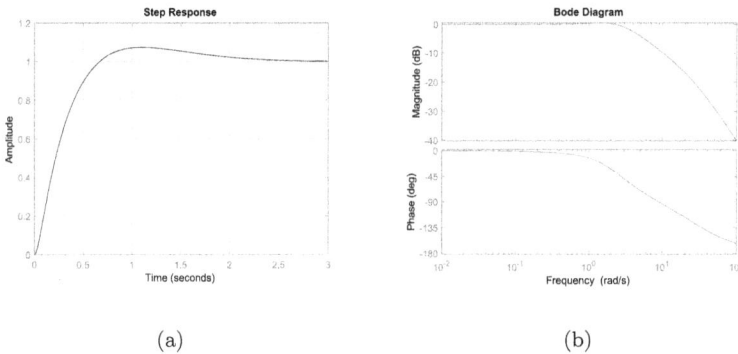

(a)

(b)

FIGURE 17.6

Behavior of the closed-loop system $F(s) = \frac{C_1(s)C_{\text{lead}}(s)G(s)}{1+C_1(s)C_{\text{lead}}(s)G(s)}$ in Exercise 17.3:
(a) step response; (b) Bode diagram.

Once designed the compensator in the s-domain, we have to select the sampling time T and pass to the z-domain. Notice that the smaller the sampling time, the smaller is the delay introduced by the sampling. In fact, this is equal to $T/2$, such that it leads to a reduction of the phase margin equal to $m_{\phi,d-t} = m_\phi - \omega_c \frac{T}{2} \frac{180}{\pi}$, where $m_{\phi,d-t}$ is the phase margin in the discrete-time domain. To ensure a maximum phase decrement of $5°$, the sampling time is selected to as $T = \frac{\pi}{180} \frac{10}{\omega_c} = 0.05$ s.

Using the transformation $s = \frac{2}{T} \frac{z-1}{z+1}$, we obtain

$$C(z) = \frac{1.4978(z - 0.8747)(z + 1)}{(z - 1)(z - 0.1095)} \tag{17.15}$$

To replicate the results concerning the compensator design in this example, the following MATLAB® commands can be used:

```
s=tf('s');
G=(s+2)/(s^2+4*s+3);
C1=1.5/s;
figure,bode(C1*G,':')
T=1.2/3.21;
alpha=12;
Clead=(1+T*s)/(1+T*s/alpha);
hold on
bode(Clead*C1*G)
K1=10^(15/20);
bode(K1*Clead*C1*G)
F=K1*Clead*C1*G/(1+K1*Clead*C1*G);
figure,step(F,'k')
w=logspace(-2,2,1001);
figure,bode(F,w)
```

Next, it is possible to calculate the step response obtained when using the digital controller. To this aim, we use the command c2d that computes a discrete-time model, with a specified sampling time, approximating the continuous-time model given as input of the function. The command uses a zero-order hold for the inputs, unless a different approximation method is specified. The commands that can be used are the following:

```
Ts=0.05
z=tf('z',Ts);
```

```
s=2/Ts*(z-1)/(z+1)
Cz=(37.84*s + 101.2)/(0.3738*s^2 + 12*s)
Gz=c2d(G,0.05)
Fz=Cz*Gz/(1+Cz*Gz)
figure,step(Fz,'k')
```

The obtained step response is shown in Fig. 17.7.

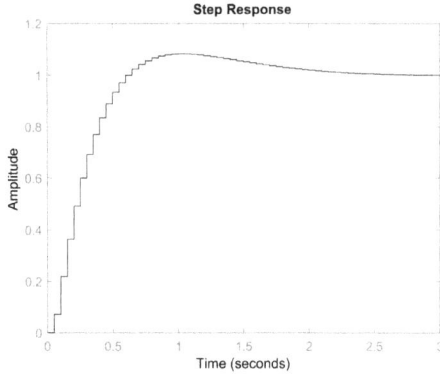

FIGURE 17.7
Step response of the feedback loop with digital control for the system in Exercise 17.3.

17.6 Exercises

1. Given the system $G(s) = \frac{10}{s+2}$, design a controller that ensures the following performance specifications:

   • null steady-state error for the step response;

   • percentage overshoot $s_\% \leq 20\%$;

   • settling time $t_{s,1\%} \leq 850$ ms.

2. Given the system $G(s) = \frac{25}{(s+5)^2}$, design a controller that ensures the following performance specifications:

   • null steady-state error for the step response;

   • steady-state error to the unit ramp, $e_{ss} \leq 0.1$;

   • percentage overshoot $s_\% \leq 20\%$;

   • rise time $t_r \leq 750$ ms.

3. Given the system $G(s) = \frac{4}{s^2+2*s+4}$, design a controller that ensures the following performance specifications:

   •null steady-state error for the response to the unit step and ramp;

   •bandwidth $\omega_c \simeq 2$ rad/s.

4. Given the system $G(s) = \frac{s/4+1}{(s+1)^2}$, design a controller that ensures the following performance specifications:

   •null steady-state error for the step response;

   •steady-state error to the unit ramp, $e_{ss} \leq 0.05$;

   •magnitude peak $M_r \leq 1.4\%$;

   •settling time $t_{s,1\%} \leq 8.34$ s.

5. Given the system $G(s) = \frac{10(s+3)}{(s+10)(s+5)}$, design a controller that ensures the following performance specifications:

   •null steady-state error for the step response;

   •percentage overshoot $s_\% \leq 10\%$;

   •rise time $t_r \leq 130$ ms.

18

Lecture 18—PID controllers

In this lecture, we illustrate the key properties of a class of widely used controllers: PID controllers. We first study the control law associated with them. Then we illustrate how these controllers can be tuned. Next, we show a few practical problems arising from the use of these controllers and how they can be mitigated. Lastly, we show some numerical examples.

18.1 The control law

PID controllers represent the most used control law in industrial control systems and the demand for PID controllers continues to grow. According to Fortune Business Insights, the global PID controller market size is estimated at USD 1.42 billion in 2021, and is expected to grow to USD 1.94 billion by 2029.

The term indicates a controller combining three different actions or modes: the proportional (P) mode, the integral (I) mode, and the derivative (D) mode. The three modes operate in parallel with each other, and the whole controller operates in a feedback loop (Fig. 18.1). As we will see in the next sections, each of the three modes can be separately *tuned* by means of a parameter regulating the intensity of the mode. This is the key for the success and widespread adoption of these controllers. In fact, tuning the controller parameters makes possible to use them for different applications. Moreover, the PID controllers can be implemented using different technologies. Over the years, mechanical, pneumatic, hydraulic, and analog electronic technologies have been used to implement PID controllers. However, today, the vast majority of new installations rely on digital electronics.

PID controllers are based on the following control law:

$$u(t) = k_P e(t) + k_I \int_0^t e(t')dt' + k_D \frac{de(t)}{dt} \qquad (18.1)$$

where k_P, k_I, and k_D are three constant parameters, representing the coefficients of the proportional, integral, and derivative modes, respectively. Under

DOI: 10.1201/9781003487289-18

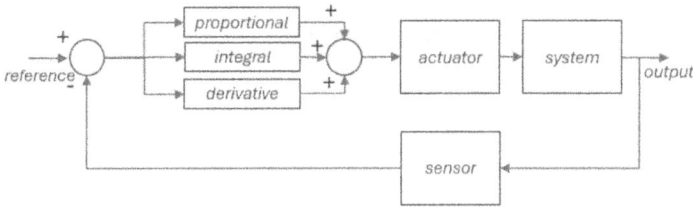

FIGURE 18.1
Feedback control loop with PID control.

the assumption that the process gain is positive, these parameters are non-negative. Each of them represents the intensity of the associated mode.

The control law (18.1) combines three terms. The purpose of the proportional mode is to correct the system behavior according to the amplitude and sign of the actual difference between set point and system output, namely the error $e(t)$. The second term, i.e., the integral mode, instead acts proportionally to the integral of the error. In this way, the control action takes into account the past history of the error, and not only its actual value. Lastly, the third term considers the derivative of the error, thus trying to anticipate the future trend of $e(t)$.

The input of the PID controller is the error $e(t)$ and its output is $u(t)$, the control signal driving the process actuator. The transfer function associated with the PID controller is:

$$C(s) = \frac{U(s)}{E(s)} = k_P + \frac{k_I}{s} + k_D s \qquad (18.2)$$

that can be rewritten as

$$C(s) = \frac{k_D s^2 + k_P s + k_I}{s} \qquad (18.3)$$

An alternative form, also called the standard form, for the PID control law is the following:

$$u(t) = k_P e(t) + \frac{k_P}{T_I} \int_0^t e(t')dt' + k_P T_D \frac{de(t)}{dt} \qquad (18.4)$$

that yields

$$u(t) = k_P \left(e(t) + \frac{1}{T_I} \int_0^t e(t')dt' + T_D \frac{de(t)}{dt} \right) \qquad (18.5)$$

where $T_I = k_I/k_P$ is the integral (or reset) time, and $T_D = k_D/k_P$ is the derivative time. In terms of transfer function, we have that

$$C(s) = \frac{U(s)}{E(s)} = k_P \left(1 + \frac{1}{T_I s} + T_D s \right) \tag{18.6}$$

The PID has two zeros $z_{1,2} = \frac{-T_I \pm \sqrt{T_I(T_I - 4T_D)}}{2T_I T_d}$ and a single pole at the origin. Therefore, it is an improper system due to the derivative term $k_D s = k_P T_D s$. In practice, the derivative mode is implemented by using $\frac{k_D s}{1 + \frac{k_D}{k_P N} s}$ or $\frac{k_P T_D s}{1 + \frac{T_D}{N} s}$, that is, by considering a further pole in $s = -N/T_D$. This further pole is added to make realizable the transfer function, ideally without influencing the performance of the controller. This means that it has to be placed far from the frequencies of interest of the control. To this aim, typically N is chosen as $N = 5 \div 10$.

Another problem of the derivative mode is the large value obtained when the set point is a step function. In fact, the derivative of $\text{step}(t)$ is always zero except at $t = 0$. Since $e(t) = r(t) - y(t)$, this yields a large value at $t = 0$, eventually moving the system out of the region of linearity. To address this issue, the diagram of Fig. 18.2 is adopted. The derivative action is obtained by acting on the system output $y(t)$ rather than on the error $e(t)$, namely by replacing $k_D \frac{de(t)}{dt}$ with $-k_D \frac{dy(t)}{dt}$ in Eq. (18.1), yielding:

$$u(t) = k_P e(t) + k_I \int_0^t e(t')dt' - k_D \frac{dy(t)}{dt} \tag{18.7}$$

or, equivalently in the standard form (18.5):

$$u(t) = k_P \left(e(t) + \frac{1}{T_I} \int_0^t e(t')dt' - T_D \frac{dy(t)}{dt} \right) \tag{18.8}$$

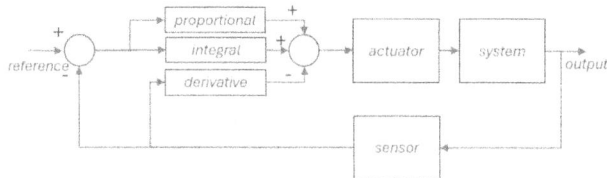

FIGURE 18.2
Feedback control loop with PID control (derivative on the output).

PID controllers do not always use all three modes. Beyond the complete case, the most typical forms are

- **P controller.** It can be obtained by setting $k_I = k_D = 0$ in Eq. (18.1) or $T_I \rightarrow \infty$ and $T_D = 0$ in Eq. (18.5). In this case we have

$$u(t) = k_P e(t) \tag{18.9}$$

and

$$C(s) = k_P \tag{18.10}$$

- **PI controller.** It can be obtained by setting $k_D = 0$ in Eq. (18.1) or $T_D = 0$ in Eq. (18.5). In this case we have

$$u(t) = k_P e(t) + k_I \int_0^t e(t')dt' = k_P \left(e(t) + \frac{1}{T_I} \int_0^t e(t')dt' \right) \tag{18.11}$$

and

$$C(s) = k_P + \frac{k_I}{s} = k_P \left(1 + \frac{1}{T_I s} \right) \tag{18.12}$$

- **PD controller.** It can be obtained by setting $k_I = 0$ in Eq. (18.1) or $T_I \rightarrow \infty$ in Eq. (18.5). In this case we have

$$u(t) = k_P e(t) + k_D \frac{de(t)}{dt} \tag{18.13}$$

and

$$C(s) = k_P + k_D s = k_P(1 + s T_D) \tag{18.14}$$

The choice between one of the possible forms is determined by the purpose and specifications of the control. For instance, if a zero steady-offset is required, then it is mandatory to include the integral action, and so a PI or a PID should be used. The derivative action suffers from the presence of noise, such that in the presence of large noise the derivative mode should be avoided. On the contrary, since the derivative mode is beneficial for stability (as it is associated with a further zero), then its use is encouraged to increase the stability margin.

18.2   PID controller tuning

The three parameters of the PID controllers, k_P, k_I, and k_D (or k_P, T_I, and T_D) determine the behavior of the closed-loop system. Although these parameters can be empirically selected, such a procedure for their choice would result in large time and costs. For this reason, over the years several techniques for the automatic *tuning* of these parameters have been developed.

In this lecture, we will only discuss one of these techniques, which is due to Ziegler and Nichols known as the Ziegler-Nichols closed-loop tuning. The first step in this procedure is to evaluate the closed-loop system behavior when only the proportional control is used. More specifically, the idea is to calculate the largest value of the gain k_P guaranteeing that the system is still stable. This value is called the ultimate gain, K_u, and corresponds to the gain margin of the open-loop system $G(s)$ (here $G(s)$ includes all blocks determining the open-loop response). Using this value of the gain, the closed-loop system will exhibit steady-state oscillations with a period $T_u = 2\pi/\omega_\pi$, where, ω_π is the frequency such that $\angle G(j\omega_\pi) = -180°$. Correspondingly, the Nyquist diagram of $G(s)$ crosses the x-axis at the point $G(j\omega_\pi) = -1/K_u$. Once known K_u and T_u, the Ziegler and Nichols method consists of selecting the tuning parameters as in Table 18.1, if the PID form in Eq. (18.1) is used, or as in Table 18.2, if the PID standard form (18.5) is used.

The Ziegler and Nichols table offers empirical rules for PID tuning in three cases: P, PI, and PID control. The controller gain k_P is selected as $K_u/2$ when the P controller is used. This means to have a stability gain margin of 2. When also the integral action is used, this margin may not be sufficient, and, for this reason, Ziegler and Nichols suggest to divide K_u by a slightly larger factor ($K_u/2.2$). On the contrary, since the derivative action is beneficial for stability, when the complete PID has to be tuned, the controller gain can be pushed closer to the ultimate value K_u, that is, k_P is tuned using $K_u/1.7$. For what concerns the choice of T_I and T_D (or analogously k_I and k_D), they are set such that $T_I = 4T_D$. Keeping in mind that the PID zeros are given by $z_{1,2} = \frac{-T_I \pm \sqrt{T_I(T_I - 4T_D)}}{2T_I T_d}$, this means to select them to be coincident.

TABLE 18.1

Values of the tuning parameters according to the Ziegler-Nichols closed-loop tuning method.

Controller	k_P	k_I	k_D
P	$K_u/2$	-	-
PI	$K_u/2.2$	$0.54K_u/T_u$	-
PID	$K_u/1.7$	$1.2K_u/T_u$	$0.075K_u T_u$

TABLE 18.2
Values of the tuning parameters according to the Ziegler-Nichols closed-loop tuning method (standard form).

Controller	k_P	T_I	T_D
P	$K_u/2$	-	-
PI	$K_u/2.2$	$T_u/1.2$	-
PID	$K_u/1.7$	$T_u/2$	$T_u/8$

MATLAB® exercise 18.1 _____

In this MATLAB® exercise we apply the Ziegler-Nichols closed-loop tuning technique to the system $G(s) = \frac{10}{(s+2)^2(s+5)}$.
The first step is to define the transfer function of the system to control, using the commands:

```
s=tf('s');
G=10/(s+2)^2/(s+5)
```

Next we calculate K_u and T_u.

```
[Ku,Pm,wc]=margin(G);
Pu=2*pi/wc;
```

We get $P_u = 1.2823$ and $K_u = 19.6074$.
Then, the controller parameters and their transfer functions are calculated.

```
%P-controller
kc_P=Ku/2;
Gc_P=kc_P;
%PI-controller
kc_PI=Ku/2.2;
Ti_PI=Pu/1.2;
Gc_PI=kc_PI*(1+1/Ti_PI/s);
%PID-controller
kc_PID=Ku/1.7;
Ti_PID=Pu/2;
Td_PID=Pu/8;
Gc_PID=kc_PI*(1+1/Ti_PI/s+s*Td_PID);
```

The values obtained are illustrated in Table 18.3.

TABLE 18.3
Ziegler-Nichols tuning (standard form) for $G(s) = \frac{10}{(s+2)^2(s+5)}$.

Controller	k_P	T_I	T_D
P	9.8037	-	-
PI	8.9125	1.0686	-
PID	11.5338	0.6412	0.1603

Using the **feedback** command we obtain the closed-loop transfer functions for the three cases, namely P control, PI control, and PID control.

```
Wcl_P=feedback(Gc_P*G,1);
Wcl_PI=feedback(Gc_PI*G,1);
Wcl_PID=feedback(Gc_PID*G,1);
```

Next, we calculate and plot the closed-loop step response for the three cases:

```
dt=0.01;
timevec=[0:dt:20];
y_P=step(Wcl_P,timevec);
y_PI=step(Wcl_PI,timevec);
y_PID=step(Wcl_PID,timevec);
figure,plot(timevec,y_P)
hold on
plot(timevec,y_PI,'r')
plot(timevec,y_PID,'k')
legend('P','PI','PID')
xlabel('t'), ylabel('y(t)')
```

Lastly, we compare the results in terms of the integral absolute error (IAE), which is defined as follows:

$$IAE = \int_0^\infty |r(t) - y(t)| dt \qquad (18.15)$$

where $r(t)$ is the reference or set point, and y the output of the closed-loop system. To compute the IAE, we use these commands:

```
SP=1;
IAE_P=dt*sum(abs(SP-y_P))
IAE_PI=dt*sum(abs(SP-y_PI))
IAE_PID=dt*sum(abs(SP-y_PID))
```

As shown in Fig. 18.3, the P control has a quite large offset, the PI control achieves zero offset by oscillations slowly damp out, the PID control represents the best case as it quickly reaches the steady-state. This is confirmed by the values of the IAE: $IAE_P = 3.8010$, $IAE_{PI} = 2.0562$, $IAE_{PID} = 0.5126$. Notice that, for the proportional controller, the improper integral involved in the definition of the IAE does not converge

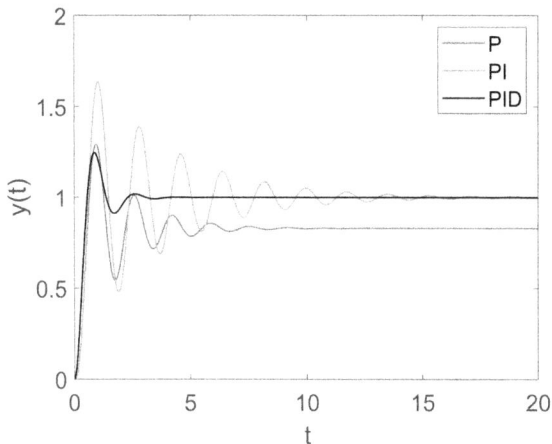

FIGURE 18.3
Example of PID control of $G(s) = \frac{10}{(s+2)^2(s+5)}$. The three controllers (P, PI, and PID) are tuned according to the Ziegler-Nichols closed-loop tuning method.

as the offset is not zero. The value calculated in MATLAB® is finite as the time window is finite.

18.3 Analytical tuning of the PID controller

In this section, an analytical procedure for tuning the PID controller is illustrated with reference to a system to control that is second-order. The procedure can be easily adapted to systems with other characteristics, i.e., transfer functions, provided that the form of the transfer function remains simple enough to perform the calculations underlined in what follows.

Let us consider the following system to control:

$$G(s) = \frac{\mu}{(1 + T_1 s)(1 + T_2 s)} \tag{18.16}$$

that is, a second-order system with two real poles, and let us consider the transfer function of a PID in standard form:

$$C(s) = k_P \left(1 + \frac{1}{sT_I} + sT_D \right) = \frac{k_P(1 + sT_I + s^2 T_I T_D)}{sT_I} \tag{18.17}$$

Using Eqs. (18.16) and (18.17), let us calculate the complementary sensitivity function $F(s)$:

$$F(s) = \frac{C(s)G(s)}{1 + C(s)G(s)} = \frac{\frac{k_P(1+sT_I+s^2 T_I T_D)}{sT_I} \frac{\mu}{(1+T_1 s)(1+T_2 s)}}{1 + \frac{k_P(1+sT_I+s^2 T_I T_D)}{sT_I} \frac{\mu}{(1+T_1 s)(1+T_2 s)}} \tag{18.18}$$

By developing the calculations, we get

$$F(s) = \frac{\frac{\mu k_P}{T_I T_1 T_2}(1 + sT_I + s^2 T_I T_D)}{s^3 + s^2 \left(\frac{1}{T_1} + \frac{1}{T_2} + \frac{\mu k_P T_D}{T_1 T_2} \right) + s \left(\frac{\mu K_P}{T_1 T_2} + \frac{1}{T_1 T_2} \right) + \frac{\mu K_P}{T_I T_1 T_2}} \tag{18.19}$$

Now, we compare the denominator of Eq. (18.19) with the one of a transfer function having two dominant complex and conjugate poles, namely having a denominator $D(s) = (s + \alpha w_n)(s^2 + 2\xi w_n s + w_n^2)$. Developing the calculations for $D(s)$ one obtains

$$D(s) = s^3 + (2\xi w_n + \alpha w_n)s^2 + (w_n^2 + 2\alpha \xi w_n^2)s + \alpha w_n^3 \tag{18.20}$$

Equating the coefficients of $D(s)$ in Eq. (18.20) and the denominator of Eq. (18.19), we obtain

$$
\begin{aligned}
\frac{\mu k_P}{T_I T_1 T_2} &= \alpha \omega_n^3 \\
\frac{\mu k_P}{T_1 T_2} + \frac{1}{T_1 T_2} &= \omega_n^2 + 2\alpha \xi \omega_n^2 \\
\frac{1}{T_1} + \frac{1}{T_2} + \frac{\mu k_P T_D}{T_1 T_2} &= 2\xi \omega_n + \alpha \omega_n
\end{aligned}
\tag{18.21}
$$

The second of Eqs. (18.21) is a function of k_P, but not of the other parameters of the PID controller, i.e., T_I and T_D. Hence, we can use it to derive k_P as follows:

$$
k_P = \frac{T_1 T_2 \omega_n^2 (1 + 2\alpha\xi) - 1}{\mu}
\tag{18.22}
$$

Then, we derive T_I and T_D from the first and the third of Eqs. (18.21), respectively, and replace the expression of k_P found in Eq. (18.22), to obtain

$$
T_I = \frac{T_1 T_2 \omega_n^2 (1 + 2\alpha\xi) - 1}{T_1 T_2 \alpha \omega_n^3}
\tag{18.23}
$$

and

$$
T_D = \frac{T_1 T_2 (2\xi\omega_n + \alpha\omega_n) - T_1 - T_2}{T_1 T_2 \omega_n^2 (1 + 2\alpha\xi) - 1}
\tag{18.24}
$$

Eqs. (18.22), (18.23), and (18.24) express the three design parameters of the PID controller as a function of the system parameters, namely μ, T_1, and T_2, and the parameters characterizing the dominant poles ξ and ω_n and the distance of the third pole from the two dominant ones, i.e., α. They can, then, be used to analytically find the parameters of the PID controller, given the system dynamics and the desired dominant poles.

MATLAB® exercise 18.2 _____

Consider $C(s)$ as in Eq. (18.17) with $\mu = 2$, $T_1 = 0.5$, and $T_2 = 3$, and tune a PID controller such that the closed-loop system has two dominant poles with $\xi = 0.4$ and $\omega_n = 2$.

First of all, we fix $\alpha = 5$ such that the third pole is far enough from the two dominant ones. Then, we apply Eqs. (18.22), (18.23), and (18.24) to find the PID parameters. We proceed as follows.

We define the system to control and the values of the parameters of the dominant poles with the commands:

```
s=tf('s')
mu=2;
T1=0.5;
T2=3;
G=mu/((1+T1*s)*(1+T2*s))
```

and

```
alpha=5;
wn=2;
xi=0.4;
```

Then, we calculate the PID controller parameters and transfer function:

```
kP=(T1*T2*wn^2*(1+2*xi*alpha)-1)/mu
Ti=(T1*T2*wn^2*(1+2*alpha*xi)-1)/(alpha*wn^3*T1*T2)
Td=(T1*T2*(2*xi*wn+alpha*wn)-T1-T2)/(T1*T2*wn^2*(1+2*xi*alpha)-1)
C=kP*(1+1/s/Ti+s*Td)
```

We obtain: k_P = 14.5, T_I = 0.4833, T_D = 0.4793 and, hence, $C(s)$ = $\frac{3.359s^2+7.008s+14.5}{0.4833s}$.

Next, we calculate $F(s)$ and compute the overshoot, the rise time and the settling time of the closed-loop step response:

```
F=feedback(C*G,1)
stepinfo(F,'SettlingTimeThreshold',0.01)
```

We obtain $s_\% = 11\%$, $t_r = 0.23$ s, and $t_{s,2\%} = 3.36$ s. Notice that these values are all smaller than those we get by using the formulas in Lecture 10 (in more detail, Eq. (10.25) for the overshoot, Eq. (10.23) for the rise time, and $t_{s,2\%} \simeq \ln(0.02)/(\xi\omega_n)$ for the settling time), namely, $s_\% = 25\%$, $t_r = 0.9$ s, and $t_{s,2\%} = 4.89$ s. This is due to the effect of the additional pole and zeros in the closed-loop system.

18.4   The integral windup problem

All actuators have physical limits and, when the controller output exceeds them, the desired motion cannot be properly realized. In the presence of actuator saturation, the integral mode can lead to the so called integral windup, or reset windup. To illustrate the problem, let us consider the system $G(s) = \frac{1}{5s+1}$ controlled by a PI with $k_P = 0.8$ and $T_I = 0.2$. Suppose that the actuator limits are ± 2.5 (see also Fig. 18.4).

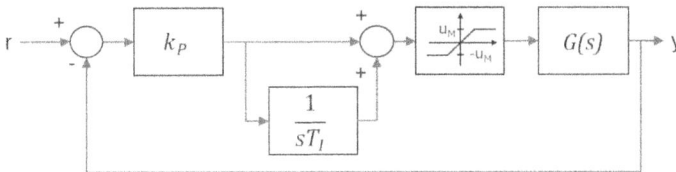

FIGURE 18.4
PI control in presence of actuator saturation.

Fig. 18.5 illustrates the closed-loop system response to $r(t) = \text{step}(t - 5)$, showing a large overshoot due to the integral windup. By inspecting the behavior of the manipulated variable (the controller output) before, $u(t)$, and after, $m(t)$, the actuator saturation (Fig. 18.6(a)), it becomes clear that desaturating the variable $u(t)$ requires a long time that deteriorates the system performance.

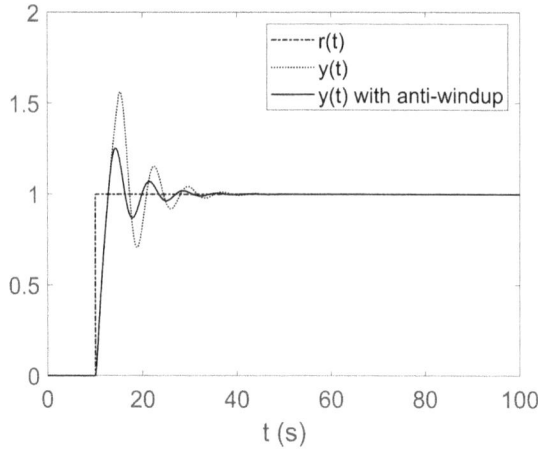

FIGURE 18.5
Closed-loop response to a step input with and without anti-windup.

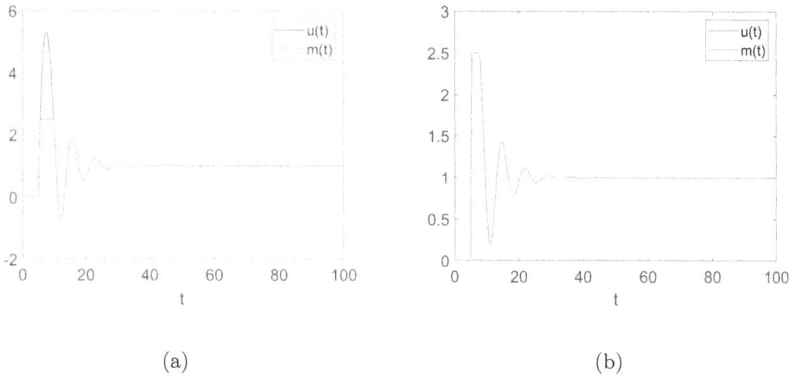

(a)   (b)

FIGURE 18.6
Manipulated variables (a) without anti-windup; (b) with anti-windup.

A possible solution to the integral windup is the scheme of Fig. 18.7 where the PI controller now includes a block modeling the actuator saturation in the direct chain of a positive feedback loop where a transfer function $\frac{1}{1+sT_I}$ is in the indirect chain. When the actuator operates in the region of linearity it can be replaced by a unit block. The transfer function of the controller, which is formed by the cascade of the constant k_P and the positive feedback loop,

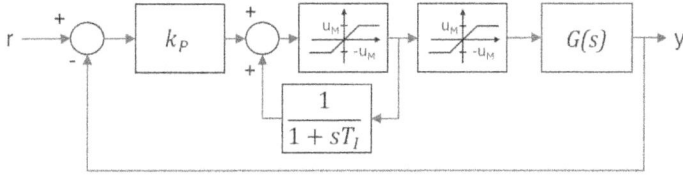

FIGURE 18.7
Anti-windup scheme.

is given by

$$C(s) = k_P \frac{1}{1 - \frac{1}{1+sT_i}} = k_P \left(1 + \frac{1}{sT_I} \right) \tag{18.25}$$

This is exactly the PI controller transfer function. On the contrary, when the variable $u(t)$ saturates, it is forced to not exceed the limits, and at the same time also the output of the block in the indirect chain converges toward the saturated value as a first order system with time scale given by T_I. Overall, this makes the controller ready to react when the error changes sign. The effect on the system output is appreciated in Fig. 18.5 where we observe a smaller overshoot compared to when the antiwindup solution is not adopted. In this case, the manipulated variable before, $u(t)$, and after, $m(t)$, the actuator saturation have exactly the same value (Fig. 18.6(b)).

18.5 Digital implementation of the PID controller

Digital implementations of PID controllers are commonly used in process control, industrial applications, robotics and many other fields. Nowadays, all new installations of PID controllers are mostly based on digital technology. In this section, we discuss the straightforward implementation of the PID control law in terms of a discrete-time system.

Let ΔT indicate the sampling interval, then for the proportional action we have

$$u_P(h) = k_P e(h) \tag{18.26}$$

For the integral action, we have

$$u_I(h) = k_I \Delta T \sum_{i=1}^{h} e(i) \tag{18.27}$$

that can be rewritten in a more computationally convenient form as:

$$u_I(h) = u_I(h-1) + k_I \Delta T e(h) \tag{18.28}$$

in such a way to avoid the calculation of the sum of many terms. This is possible by keeping memory of a variable that takes into account the result of the previous calculations (i.e., those at step $h-1$) for the integral action. Lastly, for the derivative action, we have

$$u_D(h) = k_D \frac{e(h) - e(h-1)}{\Delta T} \tag{18.29}$$

The overall control law is given by considering altogether the three contributions:

$$
\begin{aligned}
u_{PID}(h) \;=\; & u_{PID}(h-1) + k_P(e(h) - e(h-1)) + k_I \Delta T e(h) \\
& + k_D \frac{e(h) - 2e(h-1) + e(h-2)}{\Delta T}
\end{aligned} \tag{18.30}
$$

This form is particularly easy to implement in any digital processor or microcontroller. In fact, the value of the control action at any time h can be iteratively calculated after sensor reading and computation of the error $e(h)$ between reference and output. In doing this, it should be taken into account that the sampling interval ΔT represents at the same time the interval at which successive measures by the sensors of the control system are made available and the time allowed to the digital processor or microcontroller to perform all needed calculations. In the case of simple control laws such as the PID, the first time usually dominates over the second one.

18.6   Exercises

1. Using the Ziegler-Nichols method, design a PI and a PID controller for the system $G(s) = \frac{2-s}{s^2+3s+3}$ and evaluate the obtained performance.

2. Using the Ziegler-Nichols method, design a PI and a PID controller for the system $G(s) = \frac{s+2}{(s+5)^4}$ and evaluate the obtained performance.

3. Using the Ziegler-Nichols method, design a PI and a PID controller for the system $G(s) = \frac{2}{s+3} e^{-s}$ and evaluate the obtained performance.

4. Given $G(s) = \frac{1.7}{(1+0.3s)(1+2s)}$, using the method described in Sec. 18.3, find the parameters of a PID controller ensuring that the closed-loop system has $t_r \leq 0.36$ s and $s_\% \leq 20\%$, and evaluate the obtained performance.

5. Assuming that the actuator controlling the system $G(s) = \frac{10}{(s+2)^3}$ has a saturation at ± 2, design a PI controller with anti-windup, and evaluate the obtained performance.

19

Lecture 19—Linear state regulator and linear observer

The techniques for the synthesis of the controller that we have discussed in the previous lectures are based on a description of the system to control in terms of its transfer function. In this lecture, instead, we focus on an important technique that leverages the state-space representation of the system to control. This technique aims at assigning desired values to the system eigenvalues. The core of the technique is based on the feedback of all state variables, but we will also see that, under specific hypothesis, the system state variables can be *observed* from measurements of the input and output of the system, such that the obtained controller, made by the *linear regulator* and the *linear observer*, is effectively based on output feedback. This controller is of the same order, n, of the plant to control.

19.1   The linear state regulator

So far we have considered control systems based on output feedback. The linear state regulator, instead, makes use of feedback of all state variables. The difference between output and state feedback is illustrated in Fig. 19.1. Rather than using the information on the system output, in the case of state feedback, the controller determines the control action based on full information about the system state. As we will show in this lecture, under certain assumptions about the system, this broader information allows for full control of the system dynamics using a static controller, namely a control law that is a linear combination of the state variables. Compared to the design of controllers based on the frequency response, the linear state regulator does not require that the open-loop system is stable and minimum phase.

Let us consider a SISO LTI system in state-space form:

$$\dot{\mathbf{x}} = \mathbf{A}\mathbf{x} + \mathbf{B}u$$
$$y = \mathbf{C}\mathbf{x} + \mathbf{D}u \tag{19.1}$$

The linear state regulator is defined as follows:

$$u(t) = -\mathbf{K}\mathbf{x}(t) + r(t) \tag{19.2}$$

DOI: 10.1201/9781003487289-19

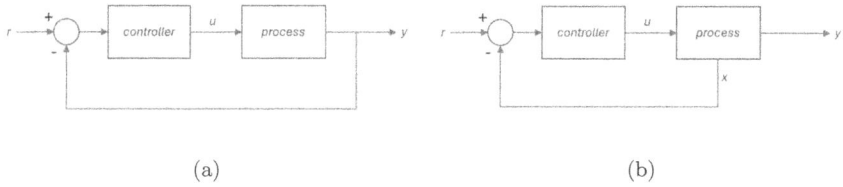

(a)   (b)

FIGURE 19.1
Closed-loop control based on (a) output or (b) state feedback.

where $K = \begin{bmatrix} k_n & k_{n-1} & \cdots & k_1 \end{bmatrix}$ is a row vector called the gain vector. The elements of K are labeled in reverse order in order to simplify the formulas that are derived below. The block diagram of the linear state regulator is illustrated in Fig. 19.2, which shows how the input of the system is formed by subtracting from the reference signal the term $-Kx(t)$.

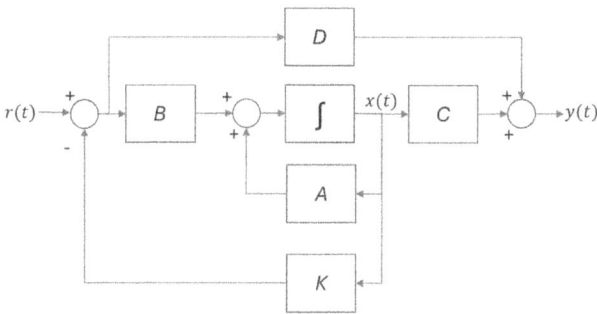

FIGURE 19.2
Block diagram of the linear state regulator.

Replacing (19.2) in the first of Eqs. (19.1), we get:

$$\dot{x} = (A - BK)x + Br \qquad (19.3)$$

The matrix $A_c = (A - BK)$ is fundamental as it shapes the dynamics of the closed-loop system. The relevant result is that, if the system is completely controllable, then the eigenvalues of A_c can be arbitrarily fixed with a proper choice of the elements of K.

To show this, let us first consider a system in reachability canonical form:

$$
A = \begin{bmatrix} 0 & 1 & 0 & \cdots & 0 \\ 0 & 0 & 1 & \cdots & 0 \\ \vdots & & & & \vdots \\ 0 & 0 & 0 & \cdots & 1 \\ -\alpha_n & -\alpha_{n-1} & -\alpha_{n-2} & \cdots & -\alpha_1 \end{bmatrix}; \quad B = \begin{bmatrix} 0 \\ 0 \\ \vdots \\ 0 \\ 1 \end{bmatrix};
$$
$$
C = \begin{bmatrix} b_n & b_{n-1} & b_{n-2} & \cdots & b_1 \end{bmatrix}; \qquad D = d
$$

(19.4)

In this case, the matrix A_c becomes

$$
A = \begin{bmatrix} 0 & 1 & 0 & \cdots & 0 \\ 0 & 0 & 1 & \cdots & 0 \\ \vdots & & & & \vdots \\ 0 & 0 & 0 & \cdots & 1 \\ -\alpha_n - k_n & -\alpha_{n-1} - k_{n-1} & -\alpha_{n-2} - k_{n-2} & \cdots & -\alpha_1 - k_1 \end{bmatrix}
$$

(19.5)

whose characteristic polynomial is $\phi_{A_c}(\lambda) = \lambda^n + (\alpha_1 + k_1)\lambda^{n-1} + \ldots + (\alpha_{n-1} + k_{n-1})\lambda + \alpha_n + k_n$.

Suppose that the desired closed-loop polynomial is given by: $\phi_d(\lambda) = \lambda^n + \phi_1\lambda^{n-1} + \ldots + \phi_{n-1}\lambda + \phi_n$. Then the coefficients of the gain vector K can be found by equating the two polynomials $\phi_{A_c}(\lambda)$ and $\phi_d(\lambda)$. This gives

$$
\begin{aligned} k_1 &= \phi_1 - a_1 \\ k_2 &= \phi_2 - a_2 \\ &\cdots \\ k_n &= \phi_n - a_n \end{aligned}
$$

(19.6)

Example 19.1 _____

Consider the following third-order SISO LTI system in reachability canonical form:

$$
A = \begin{bmatrix} 0 & 1 & 0 \\ 0 & 0 & 1 \\ -3 & 2 & 1 \end{bmatrix}; \quad B = \begin{bmatrix} 0 \\ 0 \\ 1 \end{bmatrix};
$$
$$
C = \begin{bmatrix} 1 & -1 & 2 \end{bmatrix}; \quad D = 0
$$

(19.7)

The characteristic polynomial of A is $\phi_A(\lambda) = \lambda^3 - \lambda^2 - 2\lambda + 3$. The system is unstable. The goal of the control is to make it stable, placing the closed-loop eigenvalues in $\lambda_1 = -1$, $\lambda_2 = -5$, and $\lambda_3 = -10$.
The target characteristic polynomial is $\phi_d(\lambda) = (\lambda - \lambda_1)(\lambda - \lambda_2)(\lambda - \lambda_3) = \lambda^3 + 16\lambda^2 + 65\lambda + 50$.
By using the relationships (19.6), we find:

$$
\begin{aligned} k_1 &= 16 + 1 = 17 \\ k_2 &= 65 + 2 = 67 \\ k_3 &= 50 - 3 = 47 \end{aligned}
$$

(19.8)

Consider now the more general case of a system that is not in reachability canonical form.

As the system is completely controllable, then we can preliminarily transform it into the reachability canonical form where

$$\tilde{A} = \begin{bmatrix} 0 & 1 & 0 & \cdots & 0 \\ 0 & 0 & 1 & \cdots & 0 \\ \vdots & & & & \vdots \\ 0 & 0 & 0 & \cdots & 1 \\ -\alpha_n & -\alpha_{n-1} & -\alpha_{n-2} & \cdots & -\alpha_1 \end{bmatrix} ; \quad \tilde{B} = \begin{bmatrix} 0 \\ 0 \\ \vdots \\ 0 \\ 1 \end{bmatrix} \tag{19.9}$$

where $\alpha_1, \alpha_2, \ldots, \alpha_n$ are the coefficients of the characteristic polynomial of A. Here, for easyness of notation, the reachability canonical form is denoted with a tilde symbol.

Once the system is in reachability canonical form, we proceed in an analogous fashion to the previous case, obtaining the gain vector $\tilde{K}$. Finally we can transform back the gain vector $\tilde{K}$ into the original representation using:

$$K = \tilde{K}T^{-1} \tag{19.10}$$

where T is the transformation matrix to the reachability canonical form that, taking into account Eq. (9.13), can be calculated as follows:

$$T = M_r \tilde{M}_r^{-1} \tag{19.11}$$

where M_r and $\tilde{M}_r$ are the controllability matrices of the original state-space representation and of the reachability canonical form, respectively. Notice that, using the transformation matrix T, we can eventually calculate the matrix $\tilde{C} = CT$, while the matrices $\tilde{A}$ and $\tilde{B}$ are instead written directly using and the knowledge of the characteristic polynomial of A.

Example 19.2

Consider the following third-order SISO LTI system:

$$A = \begin{bmatrix} 1 & 0 & -1 \\ 2 & -1 & -1 \\ 0 & 1 & 0 \end{bmatrix} ; \quad B = \begin{bmatrix} 1 \\ 0 \\ 2 \end{bmatrix} ; \tag{19.12}$$
$$C = \begin{bmatrix} 1 & 1 & 1 \end{bmatrix} ; \quad D = 0$$

and assume that the goal of the control is to place the closed-loop eigenvalues in $\lambda_1 = -1$, $\lambda_2 = -2$, and $\lambda_3 = -3$.

The controllability matrix, given by

$$M_r = \begin{bmatrix} B & AB & A^2B \end{bmatrix} = \begin{bmatrix} 1 & -1 & -1 \\ 0 & 0 & -2 \\ 2 & 0 & 0 \end{bmatrix} \tag{19.13}$$

is full rank. Hence, we can write the system in reachability canonical form. Since the characteristic polynomial of A is $\phi_A(\lambda) = \det(\lambda I - A) = \lambda^3 + 1$, then the reachability canonical form is given by

$$\tilde{A} = \begin{bmatrix} 0 & 1 & 0 \\ 0 & 0 & 1 \\ -1 & 0 & 0 \end{bmatrix} ; \quad \tilde{B} = \begin{bmatrix} 0 \\ 0 \\ 1 \end{bmatrix} \tag{19.14}$$

At this point, we calculate the gain vector $\tilde{K}$ using the relationships (19.6) and taking into account that $\phi_d(\lambda) = (\lambda - \lambda_1)(\lambda - \lambda_2)(\lambda - \lambda_3) = \lambda^3 + 6\lambda^2 + 11\lambda + 6$ and $\phi_{\tilde{A}}(\lambda) = \lambda^3 + 1$. We obtain $\tilde{K} = \begin{bmatrix} 5 & 11 & 6 \end{bmatrix}$.

Next, we calculate the controllability matrix in the reachability canonical form

$$\tilde{M}_r = \begin{bmatrix} \tilde{B} & \tilde{A}\tilde{B} & \tilde{A}^2\tilde{B} \end{bmatrix} = \begin{bmatrix} 0 & 0 & 1 \\ 0 & 1 & 0 \\ 1 & 0 & 0 \end{bmatrix} \tag{19.15}$$

and then the transform matrix T

$$T = M_r\tilde{M}_r^{-1} = \begin{bmatrix} -1 & -1 & 1 \\ -2 & 0 & 0 \\ 0 & 0 & 2 \end{bmatrix} \tag{19.16}$$

Finally, we calculate K

$$K = \tilde{K}T^{-1} = \begin{bmatrix} -11 & 3 & 8.5 \end{bmatrix} \tag{19.17}$$

To check the results, one can compute $A_c = A - BK$ and check that the closed-loop eigenvalues are the desired ones.

19.2   The state observer

The linear regulator works using the information on the full-state vector. When not all state variables can be measured, under certain hypotheses it is possible to reconstruct them. This is the purpose of the state observer, which, starting from measurements of the input and output of the system, provides the observed state variables that asymptotically converge to the system state variables. Also in this case, we illustrate the results by referring to SISO systems:

$$\begin{aligned} \dot{\mathbf{x}} &= A\mathbf{x} + Bu \\ y &= C\mathbf{x} + Du \end{aligned} \tag{19.18}$$

We assume that the system model, namely the matrices A, B, C, and D are known.

The state observer is a dynamical system defined by the following equations:

$$\begin{aligned} \dot{\hat{\mathbf{x}}} &= A\hat{\mathbf{x}} + Bu + H(y - \hat{y}) \\ \hat{y} &= C\hat{\mathbf{x}} + Du \end{aligned} \tag{19.19}$$

From these equations, we notice that the state observer is built as a copy of the system with the inclusion of a corrective term based on the difference between the estimated output $\hat{y}$ and the system output y (see also Fig. 19.3). This term depends on the column vector $H \in \mathbb{R}^n$, whose selection is crucial for the performance of the observer.

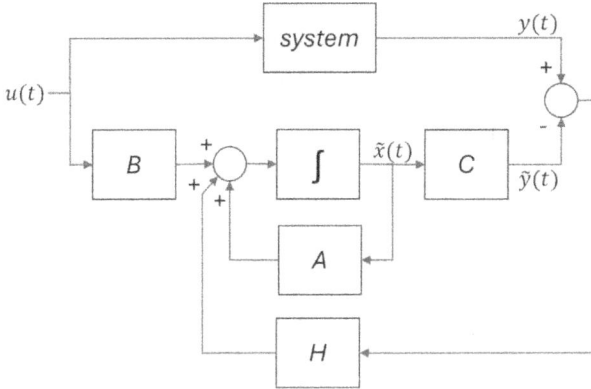

FIGURE 19.3
Block diagram of the state observer.

The vector H is called the observer gain vector and is the design parameter for the state observer. Its elements are labeled as follows:

$$H = \begin{bmatrix} h_n \\ h_{n-1} \\ \vdots \\ h_1 \end{bmatrix} \tag{19.20}$$

The choice of H directly influences the performance of the observer. To show this, we focus on the observation error $e(t) = \hat{x}(t) - x(t)$ and calculate its dynamics by subtracting Eqs. (19.18) from Eqs. (19.19):

$$\dot{e} = (A - HC)\, e \tag{19.21}$$

with initial condition $e(0) = \hat{x}(0) - x(0)$. We note that Eqs. (19.21) represent an autonomous dynamical system. The error variables, therefore, evolve in free evolution as $e(t) = e^{A_o t} e(t)$ with $A_o = A - HC$. Specifically, the error asymptotically converges to zero when A_o has all eigenvalues with a negative real part. Correspondingly, the observed variables $\hat{x}(t)$ asymptotically converge to $x(t)$. Note also that the problem becomes trivial when $e(0) = 0$; however, this condition is hard to fulfill in practice as it means that the initial condition of the system is exactly known, that is $\hat{x}(0) = x(0)$.

As mentioned above, the condition for the convergence of the observed variables to the true value of the system state variable is that A_o has all eigenvalues with a negative real part. Hence, the problem is to design the vector H such that this occurs. When the system is completely observable, this is always possible, as all eigenvalues of A_o can be arbitrarily selected.

These considerations underscore the analogy with the problem of designing the linear regulator. In fact, it is possible to proceed in full analogy with the design of the linear regulator on the dual system, and then obtain H as $H = K^T$.

An important aspect is the selection of the eigenvalues of A_o. The key condition is that they should have a negative real part. However, this may not suffice if the error dynamics converge to zero too slowly compared to the system dynamics. If the system to observe is stable, then an empirical criterion for the selection of these eigenvalues is that they should be $5 \div 10$ times smaller than the system eigenvalues. If the system is not stable, then for the eigenvalues with a positive real part the sign should be reversed when the criterion is applied, such that all eigenvalues of the observer dynamics are located in the open left half plane.

Let us illustrate the design of the state observer, considering first a system in observability canonical form:

$$A = \begin{bmatrix} 0 & 0 & \cdots & 0 & -\alpha_n \\ 1 & 0 & \cdots & 0 & -\alpha_{n-1} \\ 0 & 1 & \cdots & 0 & -\alpha_{n-2} \\ \vdots & & & & \vdots \\ 0 & 0 & \cdots & 1 & -\alpha_1 \end{bmatrix} ; \quad B = \begin{bmatrix} b_1 \\ b_2 \\ b_3 \\ \vdots \\ b_n \end{bmatrix} ;$$

$$C = \begin{bmatrix} 0 & 0 & \cdots & 0 & 1 \end{bmatrix} ; \quad D = d$$

$$(19.22)$$

By direct computation one finds that, in this case, the matrix $A_o = A - HC$ reads:

$$A_o = \begin{bmatrix} 0 & 0 & \cdots & 0 & -h_n - \alpha_n \\ 1 & 0 & \cdots & 0 & -h_{n-1} - \alpha_{n-1} \\ 0 & 1 & \cdots & 0 & -h_{n-2} - \alpha_{n-2} \\ \vdots & & & & \vdots \\ 0 & 0 & \cdots & 1 & -h_1 - \alpha_1 \end{bmatrix}$$

$$(19.23)$$

The characteristic polynomial of this matrix is $\phi_{A_o}(\lambda) = \lambda^n + (\alpha_1 + h_1)\lambda^{n-1} + \ldots + (\alpha_{n-1} + h_{n-1})\lambda + \alpha_n + h_n$.

Let $\phi_d(\lambda) = \lambda^n + \phi_1\lambda^{n-1} + \ldots + \phi_{n-1}\lambda + \phi_n$ be the desired polynomial of A_o, then to find the coefficients of the observer vector H, we equate the two polynomials $\phi_{A_o}(\lambda)$ and $\phi_d(\lambda)$, finding that the coefficients of H are given by

$$\begin{aligned} h_1 &= \phi_1 - \alpha_1 \\ h_2 &= \phi_2 - \alpha_2 \\ &\cdots \\ h_n &= \phi_n - \alpha_n \end{aligned}$$

$$(19.24)$$

Example 19.3

Let us design a state observer for the following system:

$$A = \begin{bmatrix} 0 & 0 & -30 \\ 1 & 0 & -31 \\ 0 & 1 & -10 \end{bmatrix}; \quad B = \begin{bmatrix} 0 \\ 0 \\ 1 \end{bmatrix};$$

$$C = \begin{bmatrix} 0 & 0 & 1 \end{bmatrix}; \quad D = 0 \tag{19.25}$$

The system is in observability canonical form, so it is completely observable. Moreover, the characteristic polynomial of A is $\phi_A(\lambda) = \lambda^3 + 10\lambda^2 + 31\lambda + 30$, that is rewritten as $\phi_A(\lambda) = (\lambda+2)(\lambda+3)(\lambda+5)$, showing that the system is stable.

We select the eigenvalues of A_o such that they are ten times smaller than those of A, this means that $\phi_d(\lambda) = (\lambda+20)(\lambda+30)(\lambda+50) = \lambda^3 + 100\lambda^2 + 3100\lambda + 30000$. Next, we apply the relationships (19.24), finding that $H = \begin{bmatrix} 29970 & 3069 & 90 \end{bmatrix}^T$.

We now move to the case where the system to observe is not in observability canonical form. We proceed in a way analogous to the design of the linear regulator. First, we transform the system in observability canonical form, calculate the observer gain in this representation and then the corresponding vector at the original representation. For simplicity of notation, the observability canonical form is indicated with a tilde symbol.

So, the first step is to write the matrices $\tilde{A}$ and $\tilde{C}$ of the observability canonical form:

$$\tilde{A} = \begin{bmatrix} 0 & 0 & \cdots & 0 & -\alpha_n \\ 1 & 0 & \cdots & 0 & -\alpha_{n-1} \\ 0 & 1 & \cdots & 0 & -\alpha_{n-2} \\ \vdots & & & & \vdots \\ 0 & 0 & \cdots & 1 & -\alpha_1 \end{bmatrix};$$

$$\tilde{C} = \begin{bmatrix} 0 & 0 & \cdots & 0 & 1 \end{bmatrix} \tag{19.26}$$

where $\alpha_1, \alpha_2, \ldots, \alpha_n$ are the coefficients of the characteristic polynomial of A.

The next step is to calculate the observability matrices in the two representations, that is, M_o and $\tilde{M}_o$, and the transformation matrix T from the relationship $T = M_o^{-1}\tilde{M}_o$ (see Eq. (9.29)).

Once the system has been represented in observability canonical form, we can calculate $\tilde{H}$ using the relationship (19.24), and finally compute the observer gain at the original representation as follows:

$$H = T\tilde{H} \tag{19.27}$$

Example 19.4

Consider the following system:

$$A = \begin{bmatrix} -1 & -1 & 0 \\ 1 & -1 & 2 \\ 0 & 0 & -3 \end{bmatrix}; \quad B = \begin{bmatrix} 1 \\ 1 \\ 1 \end{bmatrix};$$

$$C = \begin{bmatrix} 0 & 2 & 2 \end{bmatrix}; \quad D = 0 \tag{19.28}$$

The observability matrix is

$$M_o = \begin{bmatrix} C \\ CA \\ CA^2 \end{bmatrix} = \begin{bmatrix} 0 & 2 & 2 \\ 2 & -2 & -2 \\ -4 & 0 & 2 \end{bmatrix} \tag{19.29}$$

and it is full rank. Hence, it is possible to write the system in observability canonical form.

To this aim, we calculate the characteristic polynomial of A, which will also allow to get the system eigenvalues and understand whether the system is stable or not. We have that $\phi_A(\lambda) = \det(\lambda I - A) = \lambda^3 + 5\lambda^2 + 8\lambda + 6 = (\lambda + 3)(\lambda^2 + 2\lambda + 2)$. The system is stable. For the design of the observer we can select the desired eigenvalues ten times smaller than those of A; for instance we can select $\phi_d(\lambda) = (\lambda + 30)(\lambda + 10)^2 = \lambda^3 + 50\lambda^2 + 700\lambda + 3000$. We now write the matrices $\tilde{A}$ and $\tilde{C}$ of the observability canonical form:

$$\tilde{A} = \begin{bmatrix} 0 & 0 & -6 \\ 1 & 0 & -8 \\ 0 & 1 & -5 \end{bmatrix} ;$$
$$\tilde{C} = \begin{bmatrix} 0 & 0 & 1 \end{bmatrix}$$

(19.30)

Next, we calculate the observer vector $\tilde{H}$ using the relationships (19.24), obtaining $\tilde{H} = \begin{bmatrix} 256 & -639 & 661 \end{bmatrix}^T$.

The observability matrix $\tilde{M}_o$ is given by

$$\tilde{M}_o = \begin{bmatrix} \tilde{C} \\ \tilde{C}\tilde{A} \\ \tilde{C}\tilde{A}^2 \end{bmatrix} = \begin{bmatrix} 0 & 0 & 1 \\ 0 & 1 & -5 \\ 1 & -5 & 17 \end{bmatrix}$$

(19.31)

and so the transform matrix T is given by

$$T = M_o^{-1}\tilde{M}_o = \begin{bmatrix} 0 & 0.5 & -2 \\ -0.5 & 1.5 & -4 \\ 0.5 & -1.5 & 4.5 \end{bmatrix}$$

(19.32)

Finally, we calculate H

$$H = T\tilde{H} = \begin{bmatrix} 256 \\ -639 \\ 661 \end{bmatrix}$$

(19.33)

To check the results, one can compute $A_o = A + HC$ and check that the eigenvalues of A_o are the desired ones.

19.3   The compensator

When the system states are not-measurable, the control law of the linear regulator has to operate with the observed states, that is:

$$u(t) = -K\hat{x}(t) + r(t)$$

(19.34)

Therefore, combining together the state variables and the ones of the observer we have that

$$\dot{x} = Ax - BK\hat{x} + Br$$
$$\dot{\hat{x}} = (A - BK - HC)\hat{x} + HCx + Br$$

(19.35)

Let us now calculate the error dynamics as follows:

$$\begin{aligned}
\dot{\mathbf{e}} &= \dot{\hat{\mathbf{x}}} - \dot{\mathbf{x}} \\
&= (A - BK - HC)\,\hat{\mathbf{x}} + HC\mathbf{x} + Br - A\mathbf{x} + BK\hat{\mathbf{x}} - Br \qquad (19.36) \\
&= (A - HC)\,\mathbf{e}
\end{aligned}$$

Finally, we replace $\hat{\mathbf{x}} = \mathbf{x} + \mathbf{e}$ in the first of Eqs. (19.35), obtaining

$$\dot{\mathbf{x}} = (A - BK)\,\mathbf{x} - BK\mathbf{e} + Br \qquad (19.37)$$

We can represent the overall system formed by the system controlled by the linear regulator and the state observer using the state vector $\mathbf{z}(t) \in \mathbb{R}^{2n}$ defined as follows:

$$\mathbf{z}(t) = \left[\begin{array}{c} \mathbf{x}(t) \\ \mathbf{e}(t) \end{array} \right] \qquad (19.38)$$

In the representation with state variables $\mathbf{z}(t)$, the system equations read:

$$\dot{\mathbf{z}}(t) = \left[\begin{array}{cc} A - BK & -BK \\ 0 & A - HC \end{array} \right] \mathbf{z}(t) + \left[\begin{array}{c} B \\ 0 \end{array} \right] r(t) \qquad (19.39)$$

Let $\bar{A} = \left[\begin{array}{cc} A - BK & -BK \\ 0 & A - HC \end{array} \right]$ and $\bar{B} = \left[\begin{array}{c} B \\ 0 \end{array} \right]$, then we can rewrite the system in compact form as follows:

$$\dot{\mathbf{z}}(t) = \bar{A}\mathbf{z}(t) + \bar{B}r(t) \qquad (19.40)$$

We notice that $\bar{A}$ is block-triangular, and its eigenvalues are the union of those from the first block, $A - BK$, and those from the second block, $A - HC$. This leads to the important conclusion that the overall system is asymptotically stable if and only if the eigenvalues of $A - BK$ and $A - HC$ have negative real parts. If the system (A, B, C, D) is completely controllable and observable, then all eigenvalues can be arbitrarily selected, and, specifically, can be set so that the overall system is asymptotically stable. This can be done by selecting the two vectors K and H. Remarkably, the designs of K and H are independent of one another: the linear regulator is designed under the assumption that all system states are available for feedback, while the observer is designed as if the system were already asymptotically stable. This important result is known as the separation principle.

From a practical standpoint, designing the compensator composed of the linear state regulator and the state observer requires following the procedures outlined in Secs. 19.1 and 19.2. The eigenvalues of the closed-loop system are chosen to have negative real parts, while the eigenvalues of the state observer are selected to be faster than those of the closed-loop system, ensuring that the estimation error converges more quickly than the dynamics of the closed-loop system.

The flexibility in choosing the closed-loop eigenvalues is a significant aspect. In fact, there are techniques that leverage this flexibility of choice to incorporate additional considerations into the regulator design, such as the energy associated with the control action. This forms the basis of optimal control, an important technique that is not covered in this book.

Example 19.5 _____

Design the compensator for the LTI system with the following state matrices

$$A = \begin{bmatrix} -1 & 0 \\ 0 & 3 \end{bmatrix}; \quad B = \begin{bmatrix} 2 \\ 1 \end{bmatrix};$$
$$C = \begin{bmatrix} 1 & 2 \end{bmatrix}; \quad D = 0 \tag{19.41}$$

The system is in diagonal form, from which we can immediately conclude that it is not stable as one eigenvalue, $\lambda_2 = 3$, is positive. It is also completely controllable as all elements of B are non-zero, and completely observable as all elements of C are non-zero. Let us now select the closed-loop eigenvalues in -2 and -3, i.e., $\phi_d(\lambda) = (\lambda+2)(\lambda+3) = \lambda^2 + 5\lambda + 6$ for the linear state regulator. The eigenvalues of the state observer as selected to be ten times faster than the closed-loop eigenvalues, which yields $\phi_d(\lambda) = (\lambda + 20)(\lambda + 30) = \lambda^2 + 50\lambda + 600$ for the state observer.

At variance than Secs. 19.1 and 19.2, we now proceed to design of K and H directly on the original state representation, even if the system is not in reachability canonical form or in observability canonical form. This is still affordable as the system order is very low.

Let us calculate $\phi_{A-BK}(\lambda)$:

$$\begin{aligned} \phi_{A-BK}(\lambda) &= \det(\lambda I - A + BK) = \det \begin{bmatrix} \lambda + 1 + 2k_1 & 2k_1 \\ k_2 & \lambda - 3 + k_2 \end{bmatrix} \\ &= \lambda^2 + (2k_1 + k_2 - 2)\lambda - 6k_1 + k_2 - 3 \end{aligned} \tag{19.42}$$

Equating the coefficients of $\phi_{A-BK}(\lambda)$ with those of $\phi_d(\lambda) = \lambda^2 + 5\lambda + 6$, we get

$$\begin{aligned} 2k_1 + k_2 - 2 &= 5 \\ -6k_1 + k_2 - 3 &= 6 \end{aligned} \tag{19.43}$$

that gives: $k_1 = -0.25$, and $k_2 = 7.5$.
Next, let us calculate $\phi_{A-HC}(\lambda)$:

$$\begin{aligned} \phi_{A+HC}(\lambda) &= \det(\lambda I - A + HC) = \det \begin{bmatrix} \lambda + 1 + h_1 & 2h_1 \\ h_2 & \lambda - 3 + 2h_2 \end{bmatrix} \\ &= \lambda^2 + (h_1 + 2h_2 - 2)\lambda - 3h_1 + 2h_2 - 3 \end{aligned} \tag{19.44}$$

Equating the coefficients of $\phi_{A-HC}(\lambda)$ with those of $\phi_d(\lambda) = \lambda^2 + 50\lambda + 600$, we get

$$\begin{aligned} h_1 + 2h_2 - 2 &= 50 \\ -3h_1 + 2h_2 - 3 &= 600 \end{aligned} \tag{19.45}$$

that gives: $h_1 = -137.75$, and $h_2 = 94.875$.

19.4   MATLAB procedure for compensator design

The previous sections focused on the case of SISO systems. However, the design of both the linear state regulator and the state observer can also be extended to MIMO systems. To simplify the theoretical discussion, we concentrated on SISO systems, but, in this section, we address the MATLAB® procedure for the more general case of MIMO systems.

If both the reachability matrix and the observability matrix are full rank, then both gain matrices, K and H, can be obtained by arbitrarely assigning the eigenvalues of the matrices A_c and A_o. To this aim, the MATLAB® command `place` can be used. Assuming that the MATLAB® vectors pc and po contains the distinct eigenvalues we want to assign, then the gain matrix K can be obtained as follows:

```
K=place(A,B,pc)
```

and the gain matrix H as follows:

```
H=-place(A',B',po)'
```

In this way, the matrix A − BK will have the eigenvalues listed in the MATLAB® vector pc and the matrix A − HC will have eigenvalues listed in the MATLAB® vector po. To fix these eigenvalues, we follow the same indications of the SISO case. Specifically, we have that:

- The eigenvalues of A − BK determine the closed-loop stability. Therefore, they have to be selected in the left half-plane, considering the desired transient performance to be achieved.

- The eigenvalues of A − HC should be positioned to the left of the eigenvalues of A − BK, ensuring that the state observer converges to the state estimates faster than the control dynamics.

Also in the MIMO case, there exist analytical criteria for optimizing the placement of these eigenvalues, which belong to the theory of optimal control, not convered in this book.

19.5 MATLAB examples

In the first MATLAB® exercise, we check the results of Example 19.1.

MATLAB® exercise 19.1 _____

Let us consider the problem discussed in Example 19.1.
First, we define the matrices A and B of the system

```
A=[0 1 0; 0 0 1; -3 2 1]
B=[0 0 1]'
```

Next, we define the gain vector calculated using the relationships (19.6)

```
K=[47 67 17]
```

Finally, we calculate the closed-loop eigenvalues:

```
eig(A-B*K)
```

We obtain $\lambda_1 = -1$, $\lambda_2 = -5$, and $\lambda_3 = -10$ that correspond to the goal of the control. Alternatively, we could use the command acker, which takes as input the matrices A and B and the desired closed-loop eigenvalues, as follows

```
K=acker(A,B,[-1 -5 -10])
```

that returns the gain vector calculated above, namely $K = \begin{bmatrix} 47 & 67 & 17 \end{bmatrix}$.

In the second MATLAB® exercise, we illustrate the commands to check the results of Example 19.2.

MATLAB® exercise 19.2 _____

Let us now consider the problem illustrated in Example 19.2.
As usually, we begin with the definition of the system matrices A and B:

```
A=[1 0 -1; 2 -1 -1; 0 1 0]
B=[1 0 2]'
```

Next, we compute the reachability matrix and check if it is full rank or not:

```
Mr=ctrb(A,B)
rank(Mr)
```

Since it is full rank, the system is completely controllable and we can write it in reachability canonical form:

```
p=poly(A)
Atilde=[0 1 0; 0 0 1; -p(4) -p(3) -p(2)]
Btilde=[0 0 1]'
```

Using the relationships (19.6) we calculate the gain vector $\tilde{K}$:

```
Ktilde=[6-1 11 6]
```

and then K

```
Mrtilde=ctrb(Atilde,Btilde)
T=Mr*inv(Mrtilde)
K=Ktilde*inv(T)
eig(A-B*K)
```

Lastly, we check that the closed-loop system has the desidered eigenvalues:

```
eig(A-B*K)
```

The commands discussed above only have the purpose to illustrate the procedure, but in practice the command place, which directly computes K, should be used:

```
K=place(A,B,[-1 -2 -3])
```

In the third MATLAB® exercise, we illustrate the commands related to Example 19.3.

MATLAB® exercise 19.3

Let us consider the problem discussed in Example 19.3.
First, we define the matrices A and C of the system and calculate the system eigenvalues:

```
A=[0 0 -30; 1 0 -31; 0 1 -10]
C=[0 0 1];
eig(A)
```

Next, we define the vector H calculated using the relationships (19.24)

```
H=[29970;
   3069;
     90]
```

and check that the observer eigenvalues are the desired ones:

```
Ao=A-H*C
```

As expected, we obtain $\lambda_1 = -20$, $\lambda_2 = -30$, and $\lambda_3 = -50$.
Alternatively, we could use the command **acker**. In this case, we use as input the matrices A^T and C^T and the desired observer eigenvalues, as follows:

```
H=(acker(A',C',[-20 -30 -50]))'
```

that returns the vector H calculated above, namely $H = \begin{bmatrix} 29970 & 3069 & 90 \end{bmatrix}^T$.

In the following MATLAB® exercise, we discuss the commands related to Example 19.4.

MATLAB® exercise 19.4

Let us now consider the problem illustrated in Example 19.4.
First, we define the system matrices A and C and calculate the system eigenvalues:

```
A=[-1 -1 0; 1 -1 2; 0 0 -3]
C=[0 2 2];
eig(A)
```

Next, we compute the observability matrix and check if it is full rank or not:

```
Mo=obsv(A,C)
rank(Mo)
```

Since it is full rank, the system is completely observable and we can write it in observability canonical form:

```
p=poly(A)
Atilde=[0 1 0; 0 0 1; -p(4) -p(3) -p(2)]
Ctilde=[0 0 1];
```

Using the relationships (19.24) we calculate the vector $\tilde{H}$:

```
Htilde=[3000-6; 700-8; 50-5]
```

and then H

```
Motilde=obsv(Atilde,Ctilde)
T=inv(Mo)*Motilde
H=T*Htilde
```

Lastly, we check that the state observer has the desired eigenvalues:

```
eig(A-H*C)
```

Also in this case, the commands discussed above only have the purpose to illustrate the procedure, but in practice the command **acker** should be used:

```
H=(acker(A',C',[-10 -10 -30]))'
```

In the last MATLAB® exercise, we illustrate the commands to check the results of Example 19.5.

MATLAB® exercise 19.5 _____

Let us consider the design of the compensator as in Example 19.5.
First, we define the state matrices for system 19.41 as follows:

```
A=[-1 0; 0 3];
B=[2 1]';
C=[1 2];
D=0;
```

Next, we design the linear state regulator and the state observer, independently of each other with the commands:

```
K=place(A,B,[-2 -3])
```

and

```
H=(place(A',C',[-20 -30]))'
```

Lastly, we check that the closed-loop eigenvalues are in -2 and -3:

```
eig(A-B*K)
```

and that the eigenvalues of the state observer are in -20 and -30:

```
eig(A-H*C)
```

19.6   Exercises

1. For a system with the following state-space representation

$$
A = \begin{bmatrix} 0 & 1 & 0 \\ 0 & 0 & 1 \\ -1 & 1 & -2 \end{bmatrix} ; \quad B = \begin{bmatrix} 0 \\ 0 \\ 1 \end{bmatrix} ; \\
C = \begin{bmatrix} 1 & 0 & 0 \end{bmatrix} ; \quad D = 0
$$

(19.46)

    design a linear state regulator such that the closed-loop has two dominant eigenvalues in $\lambda_{1,2} = -1 \pm j2$.

2. For a system with the following state-space representation

$$
A = \begin{bmatrix} 0.5 & 1 \\ -1 & -3 \end{bmatrix} ; \quad B = \begin{bmatrix} 2 \\ 1 \end{bmatrix} ; \\
C = \begin{bmatrix} 1 & 1 \end{bmatrix} ; \quad D = 0
$$

(19.47)

    design a linear regulator such that the closed-loop eigenvalues are $\lambda_{1,2} = -2 \pm j3$.

3. Design the state observer for a system with the following state-space representation

$$
A = \begin{bmatrix} 0 & -12 \\ 1 & -7 \end{bmatrix} ; \quad B = \begin{bmatrix} 0.3 \\ 1 \end{bmatrix} ; \\
C = \begin{bmatrix} 0 & 1 \end{bmatrix} ; \quad D = 2
$$

(19.48)

4. Design the state observer for a system with the following state-space representation

$$A = \begin{bmatrix} -0.2 & 0.1 \\ 0.4 & -0.3 \end{bmatrix}; \quad B = \begin{bmatrix} 1 \\ -0.7 \end{bmatrix};$$
$$C = \begin{bmatrix} 0.3 & -0.2 \end{bmatrix}; \quad D = 0$$

(19.49)

5. For a system with the following state-space representation

$$A = \begin{bmatrix} \frac{4}{3} & -\frac{10}{3} \\ -\frac{5}{3} & -\frac{1}{3} \end{bmatrix}; \quad B = \begin{bmatrix} 1 \\ 0 \end{bmatrix};$$
$$C = \begin{bmatrix} 2 & -1 \end{bmatrix}; \quad D = 0$$

(19.50)

design a compensator such that the closed-loop step response has $t_r \leq 10s$ and $s_\% \leq 20\%$.

20

Lecture 20—Concluding remarks

The final lecture serves as a comprehensive summary, highlighting the key guidelines that readers should retain after completing the book. Additionally, as the closing lecture of the course, it offers suggestions to reinforce the main concepts covered, further remarking on the foundational principles of automatic control.

20.1 Concluding remarks

The book is designed to be self-contained, allowing readers to follow its content in the suggested order. Each lesson builds on the previous one, so skipping lessons may lead to gaps in understanding. This structure ensures cohesive and cumulative learning, guiding readers step-by-step through foundational to more advanced topics in automatic control.

The book's organization provides a dual focus: it first introduces the foundational elements necessary for analyzing the dynamic behavior of linear time-invariant systems (Lectures 2-14), then transitions into the principles and methods for controller design (Lectures 15-19). This structure enables readers to build a comprehensive understanding of both system behavior and control strategies, essential for mastering automatic control.

The book addresses both time-domain and frequency-domain approaches, providing a comprehensive framework for analyzing system dynamics from multiple perspectives. The two lectures on discrete-time systems serve a dual purpose: they introduce this class of systems while also covering the discretization of continuous-time systems, laying the groundwork for digital controller implementation.

The authors emphasize the importance of developing a precise mathematical model of a system, as it serves as the foundation for any type of control design or analysis. In particular, the concept of the observer is highlighted as one of the most significant advancements in automatic control theory over the past fifty years, offering essential insights and applications in both theoretical study and real-world implementations. The observer's design relies fundamentally on the availability of a mathematical model. This highlights the observer's powerful capability: it enables the exploration of a system's

DOI: 10.1201/9781003487289-20

internal states using only input and output measurements. By providing insight into otherwise inaccessible system variables, the observer has become an invaluable tool in control theory. The observer is crucial in defining the feedback compensator, as it provides estimates of internal states that are essential for feedback control. Conversely, the observer itself can be designed using feedback principles, enabling it to dynamically adjust its estimates based on the difference between system and observer outputs.

Readers are encouraged to deepen their understanding by working through the exercises provided. A particularly effective way to reinforce learning is to study the theoretical concepts and then develop MATLAB® code to implement related functions.

The lectures focus primarily on closed-loop control, emphasizing its advantages in system stability and disturbance rejection. However, open-loop control is also important, especially for tracking applications where measurements are unavailable or where measurements might negatively impact system performance. The mathematical model is fundamental also in this context, as it provides the essential framework for both closed-loop and open-loop control strategies.

The physical principles employed in this book are grounded in classical Newtonian physics, which are highly applicable in industrial contexts and robotics, spanning micro to medium and large scales. However, when it comes to nanoscale systems, the mathematical models based on Newtonian physics become inadequate, necessitating the use of quantum mechanics models and approaches. This shift in control paradigms for nanoscale systems represents a significant departure from traditional methods, highlighting the need for further study and new lectures dedicated to these advanced topics.

20.2 Plan for the study

1. Install MATLAB®.

2. After each lecture, do the related examination test.

3. Read the lecture exercises, and, before solving them, read the next lecture.

4. Solve the exercises of Lecture i and proceed to study Lecture $i + 1$.

5. After solving an exercise, write down a new exercise.

6. Go back to step 2 and do a second reading of the lectures.

7. Use MATLAB® ten times for each lecture.

8. Open the book to an arbitrary page and summarize the main results.

9. Give yourself a grade from 1 to 30. If it less than 18, go to step 2. Otherwise, you are ready for the exam!

Appendix A—Examination test

This appendix provides a final evaluation test for the students to check their knowlegde of the main concepts illustrated in the different lectures.

Lecture 1

1. The actuator is:

☐ a plant

☐ a device to supply power

☐ an algorithm

2. Feedback is:

☐ a paradigm

☐ an algorithm

☐ a plant

3. How many years ago did control begin?

☐ 100 years ago

☐ 200 years ago

☐ long time ago

4. What is a system?

☐ A device powered by electricity

DOI: 10.1201/9781003487289-A

☐ Any object with an input-output casual relationship

☐ A set of equations

5. Without sensors, the control system is

☐ a closed-loop control system

☐ an open-loop control system

☐ a feedback control system

Lecture 2

1. State-space equations are:

☐ differential equations

☐ algebraic equations

☐ parabolic equations

2. Linear time-invariant systems are:

☐ systems with fixed parameters

☐ systems where time is not included in the equations

☐ systems whose equations change in time

3. Pendulum is

☐ a mechanical system

☐ an electrical system

☐ an electro-mechanical system

4. Linear time-invariant systems

☐ can be described by a set of four matrices

☐ always have a single input and a single output

☐ are the most general form of a dynamical system

5. In the quarter car model, the input is

☐ the weight of the car

☐ the shape of the road pavement

☐ the action of the motor

Lecture 3

1. Which of the following MATLAB® commands correctly defines a vector `a1` containing all odd numbers from 3 to 13?

☐ `a1=[3:13]`

☐ `a1=[2:2:13]`

☐ `a1=[3:2:14]`

2. Which MATLAB® command can be used to define the identity matrix?

☐ `ones`

☐ `eye`

☐ `rand`

3. How can a 3 × 4 matrix of random values (with uniform distribution in $[0, 1]$) be defined in MATLAB®?

☐ `rand(3,4)`

☐ `randn(4,3)`

☐ `randn(3,4)`

4. Given a 10×10 matrix `A1` defined in MATLAB®, how can we extract the second and fourth column?

☐ `A1(:,[2:4])`

☐ `A1(:,[2 4])`

☐ `A1([2:4],:)`

5. Which MATLAB® command allows one to define LTI systems?

☐ `plot`

☐ `ss`

☐ `hist`

Lecture 4

1. The Laplace transform of $u(t) = 5$ is:

☐ $U(s) = \frac{1}{s}$

☐ $U(s) = 5$

☐ $U(s) = \frac{5}{s}$

2. The inverse Laplace transform of $Y(s) = \frac{1}{s^2}$ is:

☐ $y(t) = t$

☐ $y(t) = t^2$

☐ $y(t) = \frac{1}{t^2}$

3. The Laplace transform of $u(t) = \text{step}(t) + \text{ramp}(t)$ is:

☐ $U(s) = \frac{s^2+1}{s^2}$

☐ $U(s) = \frac{s}{s^2+1}$

☐ $U(s) = \frac{s+1}{s^2}$

4. Let $y(t)$ be the signal with Laplace transform $Y(s) = \frac{s}{s^2+1}$, then its limit $\lim_{t \to +\infty} y(t)$:

☐ is equal to zero

☐ is equal to one

☐ does not exist

5. The inverse Laplace transform of $Y(s) = \frac{e^{-sT}}{s}$ with $T > 0$ is:

☐ $y(t) = \text{ramp}(t)$

☐ $y(t) = \text{step}(t - T)$

☐ $y(t) = 1$

Lecture 5

1. The system

$$\begin{aligned} \dot{x}_1 &= tx_2 \\ \dot{x}_2 &= 2x_1 \end{aligned} \tag{A.1}$$

is:

☐ linear

☐ nonlinear

☐ linear time-varying

2. The system $\dot{x} = 5x^2 + \sin t$ is:

☐ linear time-varying

☐ non-dynamical

☐ nonlinear

3. The system

$$\begin{aligned} \dot{\mathbf{x}} &= \mathbf{Ax} + \mathbf{Bu} \\ \mathbf{y} &= \mathbf{Cx} + \mathbf{Du} \end{aligned}$$ (A.2)

with $\mathbf{D} = 0$ is

☐ strictly proper

☐ linear time-varying

☐ in free evolution

4. Let $\mathbf{A} = \begin{bmatrix} 1 & 0 \\ 0 & -1 \end{bmatrix}$ and $\mathbf{x}_0 = \begin{bmatrix} 0 \\ 1 \end{bmatrix}$, then the free evolution is such that:

☐ $\lim\limits_{t \to +\infty} \mathbf{x}(t) = \begin{bmatrix} 0 \\ 0 \end{bmatrix}$

☐ $\lim\limits_{t \to +\infty} \mathbf{x}(t) = \begin{bmatrix} 0 \\ 1 \end{bmatrix}$

☐ $\lim\limits_{t \to +\infty} \mathbf{x}(t) = \begin{bmatrix} +\infty \\ 0 \end{bmatrix}$

5. Let $\mathbf{A} = \begin{bmatrix} -1 & 0 & 0 \\ 0 & -2 & 0 \\ 0 & 0 & -3 \end{bmatrix}$ and $\mathbf{x}_0 = \begin{bmatrix} 1 \\ 1 \\ 1 \end{bmatrix}$, then the free evolution is such that:

☐ $\lim\limits_{t \to +\infty} \mathbf{x}(t) = \begin{bmatrix} 0 \\ 0 \\ 0 \end{bmatrix}$

☐ $\lim\limits_{t \to +\infty} \mathbf{x}(t) = \begin{bmatrix} 1 \\ 1 \\ 1 \end{bmatrix}$

☐ $\lim\limits_{t \to +\infty} \mathbf{x}(t) = \begin{bmatrix} +\infty \\ +\infty \\ +\infty \end{bmatrix}$

Lecture 6

1. A system with characteristic polynomial $p(\lambda) = \lambda^3 + \lambda^2 + \lambda$ is:

☐ asymptotically stable

☐ stable

☐ unstable

2. A linear time-invariant system with $A = \begin{bmatrix} -1 & 4 & 0 \\ -4 & -1 & 0 \\ 0 & 0 & -5 \end{bmatrix}$ is:

☐ asymptotically stable

☐ stable

☐ unstable

3. For which value of k is the system with characteristic polynomial $p(\lambda, k) = \lambda^3 + k\lambda^2 + \lambda^2 + 1$ asymptotically stable?

☐ $k > 0$

☐ $k > 1$

☐ $\nexists k$

4. The Routh criterion

☐ gives a necessary and sufficient condition for root location on the complex plane

☐ gives a necessary condition for root location on the complex plane

☐ gives a sufficient condition for root location on the complex plane

5. In a linear system, if an equilibrium point is asympotically stable, then

☐ all other equilibrium points are stable

☐ all other equilibrium points are asymptotically stable

☐ there can be other equilibrium points with their own stability properties

Lecture 7

1. Given an LTI system with

$$A = \begin{bmatrix} -1 & 0 & 0 \\ 0 & 0 & 0 \\ 0 & 0 & 1 \end{bmatrix}; \quad B = \begin{bmatrix} 1 \\ 1 \\ 1 \end{bmatrix}; \quad \text{(A.3)}$$

$$C = \begin{bmatrix} 0 & 1 & 1 \end{bmatrix}; \quad D = 0$$

then its transfer function is:

☐ $G(s) = \frac{s+4}{s(s+1)(s-1)}$

☐ $G(s) = \frac{2s-1}{s(s-1)}$

☐ $G(s) = \frac{1}{s}$

2. For a system with $G(s) = \frac{s}{s^2+1}$, the impulse response is:

☐ $y(t) = \sin t$

☐ $y(t) = \cos t$

☐ $y(t) = e^{-t} + e^t$

3. A system with $G(s) = \frac{s-1}{s^2-1}$ is

☐ stable

☐ asymptotically stable

☐ unstable

4. How many zeros has the transfer function $G(s) = \frac{s^2+1}{(s^2+s+1)(s+1)}$

☐ 2

☐ 0

☐ 3

5. The final value of the unit step response of a system with $G(s) = \frac{s+2}{s^2+3s+1}$ is:

☐ 0

☐ 2

☐ 1

Lecture 8

1. The system with $G(s) = \frac{1}{s+1}$

☐ can be written as a feedback scheme (Fig. 8.3) with $G_1(s) = \frac{1}{s}$ and $G_2(s) = 1$

☐ can be written as a feedback scheme (Fig. 8.3) with $G_1(s) = 1$ and $G_2(s) = \frac{1}{s}$

☐ cannot be expressed as a feedback system

2. A positive feedback configuration with $G_1(s) = \frac{1}{s+1}$ and $G_2(s) = 1$ is:

☐ stable

☐ unstable

☐ asymptotically stable

3. Given the system in Fig. A.1 which one of the following statements is true?

☐ The closed-loop system is asymptotically stable

☐ The open-loop system is third order

☐ The closed-loop system is unstable

4. For the system in Fig. A.2, let $R = 0$ and $G(s) = \frac{Y(s)}{N(s)}$, then

☐ $G(s) = \frac{1}{s+1}$

☐ $G(s) = \frac{1}{s-1}$

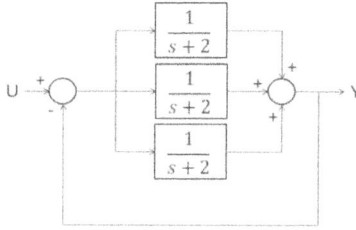

FIGURE A.1
Block scheme with a parallel of three systems inserted in a feedback configuration.

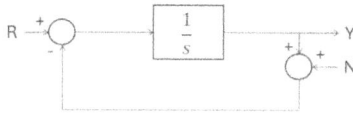

FIGURE A.2
Block scheme with a feedback configuration.

☐ $G(s) = -\frac{1}{s+1}$

5. The series connection of $G_1(s) = \frac{1}{s+1}$ and $\frac{1}{s}$ is:

☐ BIBO stable

☐ unstable

☐ stable

Lecture 9

1. A continuous-time LTI system with $B = 0$ is:

☐ completely reachable

☐ not reachable at all

☐ reachability depends on A

2. The reachable part of a system

☐ cannot be influenced by the variables of the non-reachable part

☐ can be influenced by the variables of the non-reachable part

☐ is always stable

3. The observability matrix

☐ is always square

☐ does not depend on B

☐ has always rank equal to the system order

4. A system that is not in minimal form

☐ is unreachable

☐ is unobservable

☐ has some zero-pole cancellations

5. The dual of a system with state matrices A, B, and C is the system with matrices:

☐ A^T, C^T, and B^T

☐ A^T, B^T, and C^T

☐ A^T, C, and B

Lecture 10

1. The time constant of a system with $G(s) = \frac{4}{s+2}$ is:

☐ $\tau = 2$

☐ $\tau = 0.5$

☐ $\tau = 4$

2. Which of the panels of Fig. A.3 shows the correct impulse response of the second-order system $G(s) = \frac{s+1}{(s+0.5)(s+3)}$?

☐ Fig. A.3(a)

☐ Fig. A.3(b)

☐ Fig. A.3(c)

3. Which are the dominant poles of a system with $G(s) = \frac{s+1.01}{(s+1)(s+2)}$?

☐ $s = -1$ and $s = -2$

☐ $s = -1$

☐ $s = -2$

4. Which of the following systems has an oscillatory impulse response?

☐ $G(s) = \frac{1}{s^2+2s+1}$

☐ $G(s) = \frac{1}{s^2+2s+2}$

☐ $G(s) = \frac{1}{s^s+2s+0.5}$

5. Which of the following systems has the smallest rise time t_r?

☐ $G(s) = \frac{1+s}{s^2+s+1}$

☐ $G(s) = \frac{1-s}{s^2+s+1}$

☐ $G(s) = \frac{1}{s^s+s+1}$

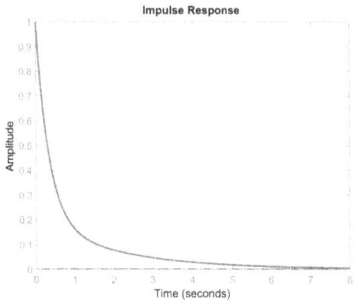

(a)

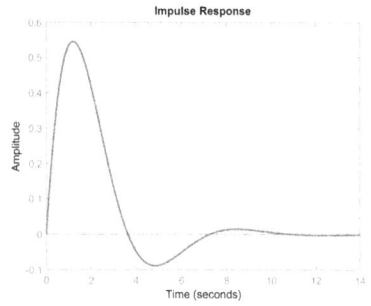

(b)

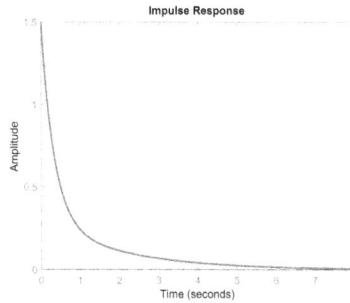

(c)

FIGURE A.3
(a)-(c) Impulse responses of three different systems.

Lecture 11

1. Let $G(s) = \frac{4}{s+2}$. For which value of ω is $|G(j\omega)|_{dB} = 0$?

☐ $\omega = 2$

☐ $\omega = 4$

☐ $\omega = 2\sqrt{3}$

2. Let $G(s) = \frac{4}{s+2}$, then $\angle G(j2)$ is equal to:

☐ $\angle G(j2) = -\pi/4$

☐ $\angle G(j2) = -\pi/2$

☐ $\angle G(j2) = -\pi$

3. The system with $G(s) = \frac{s^2+4s+4}{s^2+s+1}$ is:

☐ a low-pass filter

☐ a band-pass filter

☐ a high-pass filter

4. The system with $G(s) = \frac{(s+2)^2}{(s+4)^2}$ is:

☐ a low-pass filter

☐ a band-pass filter

☐ a high-pass filter

5. The system with the following state matrices

$$A = \begin{bmatrix} 0 & 1 \\ -1 & -1 \end{bmatrix}; \quad B = \begin{bmatrix} 0 \\ 1 \end{bmatrix};$$
$$C = \begin{bmatrix} 1 & 0 \end{bmatrix}; \quad D = 0 \tag{A.4}$$

is:

☐ a low-pass filter

☐ a band-pass filter

☐ a high-pass filter

Lecture 12

1. Which is the most significant information that can be derived from the Nyquist plot of a system?

☐ the gain margin

☐ the closed-loop stability

☐ the frequency response

2. Let $G(s) = \frac{4}{s+1}$, then the number of encirclements of the point (-1,0) by the Nyquist plot is equal to:

☐ $N = -1$

☐ $N = 0$

☐ $N = 1$

3. Let $W(s) = \frac{s+1}{s^2+s+1}$ be the closed-loop transfer function of a system, then the Nyquist plot of the corresponding open-loop system:

☐ crosses the point $(-1,0)$

☐ is such that $N = 0$

☐ is such that $N = -1$

4. Given $G(s) = \frac{1}{s^2+s+1}$, which of the following systems has a larger phase margin?

☐ $G(s) = \frac{1}{(s/10+1)(s^2+s+1)}$

☐ $G(s) = \frac{s/0.1+1}{s^2+s+1}$

☐ $G(s) = \frac{s/20+1}{(s/10+1)(s^2+s+1)}$

5. Open-loop systems with poles on the imaginary axis lead to:

☐ unstable closed-loop systems

☐ Nyquist plots with asymptotes

☐ positive phase margins

Lecture 13

1. Discrete-time systems are described by:

☐ a set of differential equations

☐ a set of finite difference equations

☐ a set of constants

2. Which of the following eigenvalues of a discrete-time LTI system yields modes with persistent oscillations?

☐ $\lambda = 0$

☐ $\lambda = -1$

☐ $\lambda_1 = 1$

3. Which of the following discrete-time LTI systems is stable?

☐ A system with $\lambda_1 = -1$ and $\lambda_1 = -2$

☐ A system with $\lambda_1 = 0$ and $\lambda_1 = -0.5$

☐ A system with $\lambda_{1,2} = -0.5 \pm j1$

4. The discrete time system with the following state matrices

$$A = \begin{bmatrix} 0 & 1 & 0 \\ 0 & 0 & 1 \\ 0 & 0 & 0 \end{bmatrix}; \quad B = \begin{bmatrix} 0 \\ 0 \\ 1 \end{bmatrix}; \quad\quad\quad \text{(A.5)}$$
$$C = \begin{bmatrix} 1 & 1 & 1 \end{bmatrix}; \quad D = 0$$

is:

☐ unstable

☐ a FIR filter

☐ a IIR filter

5. In a sample data system increasing the sampling time leads to:

☐ closed-loop instability

☐ better precision

☐ a faster transient

Lecture 14

1. The z-transform is

☐ a state-space transformation

☐ a linear operator

☐ a map from the s-plane to the z-plane

2. The bilinear transform is:

☐ a state-space transformation

☐ a frequency transformation

☐ a time transformation

3. The system with $G(z) = \frac{z+2}{z-0.5}$ is:

☐ asymptotically stable

☐ stable

☐ unstable

4. Which of the following statements holds true?

☐ An IIR can be always approximated by an FIR system of higher-order

☐ An asymptotically stable IIR system can be always approximated by an FIR system

☐ A FIR system can be always approximated by an IIR system

5. An A/D converter

☐ transforms the analog input into a digital output

☐ transforms the digital input into an analog output

☐ can be implemented using a zero-order hold

Lecture 15

1. In a closed-loop control system, the transfer function from $r(t)$ to $e(t)$ is:

☐ the sensitivity

☐ the complementary sensitivity

☐ the control sensitivity

2. Increasing the open-loop static gain is beneficial for:

☐ disturbance rejection

☐ stablity

☐ noise filtering

3. A closed-loop system with unit feedback has zero steady state error to the ramp input. How many poles at the origin has the open-loop system?

☐ 0

☐ 1

☐ 2

4. The system type is:

☐ a characteristic of the time response

☐ a characteristic of the transfer function

☐ a characteristic of the frequency response

5. While designing the steady-state compensator one must assure

☐ sensitivity performance

☐ asymptotic stability

☐ disturbance rejection

Lecture 16

1. The rise time is related to:

☐ the bandwidth

☐ the phase margin

☐ the overshoot

2. Increasing both the bandwidth and the phase margin generally requires:

☐ the use of lead compensators

☐ the use of lead-lag compensators

☐ the inclusions of further zeros in the open-loop system

3. The bandwidth can be increased by increasing:

☐ the critical frequency

☐ the settling time

☐ the phase margin

4. Lead-lag compensators are:

☐ asymptotically stable systems

☐ stable systems

☐ asymptotically stable and minimum phase systems

5. How many lead-lag networks are needed to make stable the system $G(s) = 1/s^3$?

☐ one

☐ two

☐ more than two

Lecture 18

1. PID controllers are used to control:

☐ asymptotically stable systems

☐ stable systems

☐ minimum phase systems

2. PI controllers are:

☐ type 0 systems

☐ type 1 systems

☐ type 2 systems

3. The ideal PID is:

☐ a non-strictly proper system

☐ an unstable system

☐ an universal controller

4. The derivative action is beneficial for:

☐ stability margins

☐ noise filtering

☐ steady-state error

5. A control system requires zero steady-state error to the step input. Which controllers are suitable to achieve this control goal?

☐ P, PI, PD, and PID

☐ PI and PID

☐ P and PI

Lecture 19

1. When is the use of the LQR appropriate?

☐ When the system order is high

☐ When the system is unstable

☐ When the system is discrete-time

2. An LQR can be designed provided that

☐ M_o is full rank

☐ M_c is full rank

☐ M_c and M_o are both full rank

3. The eigenvalues of A_o have to be selected:

☐ in arbitrary positions in the left half plane

☐ on the left to those of A_c

☐ on the right to those of A_c

4. The closed-loop eigenvalues are:

☐ the eigenvalues of A_c

☐ the eigenvalues of A_o

☐ the union of the eigenvalues of A_c and A_o

5. Which of the following statements on the state observer is false?

☐ The observer is in the same order of the system to reconstruct

☐ The observer can be designed provided that the system is in observable form

☐ The state variables of the observer asymptotically follow those of the system to reconstruct

Lecture 20

1. How have you tested your MATLAB® knowledge?

☐ By reproducing the commands discussed in the textbook

☐ By writing your own code

2. How have you practiced designing compensators using frequency response methods?

☐ By solving the exercises reported in the textbook

☐ By solving the exercises reported in the textbook and several other exercises

3. Is the difference between a linear state regulator and feedback control clear to you?

☐ Yes

☐ Partially

☐ No

4. Do you prefer designing a compensator using one technique over another?

☐ I prefer to use the frequency response method

☐ I prefer to use the time domain approach

☐ No preferences

5. Is the importance of the discrete-time approach and its relation to controller implementation clear to you?

☐ Yes

☐ Partially

☐ No

Selected references

Antoulas, T. (2004). Model reduction of large-scale systems. SIAM, Philadelphia.

Bhattacharyya, S. P., Datta, A., & Keel, L. H. (2018). Linear control theory: structure, robustness, and optimization. CRC Press.

Bishop, R. C., & Dorf, R. H. (2011). Modern control systems.

Bolzern, P. G. E., Scattolini, R., & Schiavoni, N. L. (2008). Fondamenti di controlli automatici. McGraw-Hill.

Brian, D. O., Anderson, & Moore, J. (2014). Optimal control. Dover Publications.

D'Azzo, J. J., & Houpis, C. D. (1995). Linear control system analysis and design: conventional and modern. McGraw-Hill Higher Education.

Duan, G.-R., & Yu, H.-H. (2013). LMIs in control systems: analysis, design and applications. CRC Press.

Engelberg, S. (2024). Mathematical introduction to control theory. World Scientific.

Fortuna, L., Frasca, M., & Buscarino, A. (2021). Optimal and robust control: advanced topics with MATLAB. CRC Press.

Franklin, G. F., Powell, J. D., Emami-Naeini, A., & Powell, J. D. (2002). Feedback control of dynamic systems. Upper Saddle River: Prentice Hall.

Horowitz, I. M. (2013). Synthesis of feedback systems. Elsevier.

Kharitonov, V. (2012). Time-delay systems: Lyapunov functionals and matrices. Springer Science & Business Media.

Knowles, J. B. (1964). Book Review: Theory of Automatic Control.

Levine, W. S. (2018). The Control Handbook. CRC Press.

Mastellone, S., & Van Delft, A. (Eds.). (2023). The impact of automatic control research on industrial innovation: enabling a sustainable future. John Wiley & Sons.

Raven, F. H. (1995). Automatic control engineering. McGraw-Hill, Inc.

Reference Data for Radio Engineers IIT. (1974).

Steinway, W. J. (1971). Estimation Theory with Applications to Communication and Control - Andrew P. Sage and James L. Melsa. (New York: McGraw-Hill, 529 pp.).

Trentelman, H. L., Stoorvogel, A. A., Hautus, M., & Dewell, L. (2002). Control theory for linear systems. Applied Mechanics Reviews, 55(5), B87-B87.

Tsui, C.-C. (2022). Robust control system design: advanced state space techniques. CRC Press.

Zadeh, L., & Desoer, C. (2008). Linear system theory: the state space approach. Courier Dover Publications.

Index

For Product Safety Concerns and Information please contact our EU
representative GPSR@taylorandfrancis.com
Taylor & Francis Verlag GmbH, Kaufingerstraße 24, 80331 München, Germany

www.ingramcontent.com/pod-product-compliance
Lightning Source LLC
Chambersburg PA
CBHW060345220326
41598CB00023B/2814

9 781032 783086